Radioactivity

STEM Road Map
for High School

**Grade
11**

Radioactivity

Grade 11

STEM Road Map
for High School

Edited by Carla C. Johnson, Janet B. Walton, and
Erin Peters-Burton

NSTApress
National Science Teachers Association

Arlington, Virginia

National Science Teachers Association

Claire Reinburg, Director
Rachel Ledbetter, Managing Editor
Andrea Silen, Associate Editor
Jennifer Thompson, Associate Editor
Donna Yudkin, Book Acquisitions Manager

ART AND DESIGN
Will Thomas Jr., Director, cover and
    interior design
Himabindu Bichali, Graphic Designer, interior
    design

PRINTING AND PRODUCTION
Catherine Lorrain, Director

NATIONAL SCIENCE TEACHERS ASSOCIATION
David L. Evans, Executive Director

1840 Wilson Blvd., Arlington, VA 22201
www.nsta.org/store
For customer service inquiries, please call 800-277-5300.

SUSTAINABLE FORESTRY INITIATIVE
Certified Chain of Custody
At Least 10% Certified Forest Content
www.sfiprogram.org
SFI-01028

Copyright © 2019 by the National Science Teachers Association.
All rights reserved. Printed in the United States of America.
22 21 20 19   4 3 2 1

NSTA is committed to publishing material that promotes the best in inquiry-based science education. However, conditions of actual use may vary, and the safety procedures and practices described in this book are intended to serve only as a guide. Additional precautionary measures may be required. NSTA and the authors do not warrant or represent that the procedures and practices in this book meet any safety code or standard of federal, state, or local regulations. NSTA and the authors disclaim any liability for personal injury or damage to property arising out of or relating to the use of this book, including any of the recommendations, instructions, or materials contained therein.

PERMISSIONS
Book purchasers may photocopy, print, or e-mail up to five copies of an NSTA book chapter for personal use only; this does not include display or promotional use. Elementary, middle, and high school teachers may reproduce forms, sample documents, and single NSTA book chapters needed for classroom or noncommercial, professional-development use only. E-book buyers may download files to multiple personal devices but are prohibited from posting the files to third-party servers or websites, or from passing files to non-buyers. For additional permission to photocopy or use material electronically from this NSTA Press book, please contact the Copyright Clearance Center (CCC) (*www.copyright.com*; 978-750-8400). Please access *www.nsta.org/permissions* for further information about NSTA's rights and permissions policies.

Library of Congress Cataloging-in-Publication Data
Names: Johnson, Carla C., 1969- editor. | Walton, Janet B., 1968- editor. | Peters-Burton, Erin E., editor.
Title: Radioactivity, grade 11 : STEM road map for high school / edited by Carla C. Johnson, Janet B. Walton, and
    Erin Peters-Burton.
Description: Arlington, VA : National Science Teachers Association, [2019] | Includes bibliographical references
    and index.
Identifiers: LCCN 2018046668 (print) | LCCN 2018049965 (ebook) | ISBN 9781681404752 (e-book) |
    ISBN 9781681404745 (print)
Subjects: LCSH: Radiation--Study and teaching (Secondary) | Radioactive pollution--Environmental aspects--
    Study and teaching (Secondary) | Eleventh grade (Education)
Classification: LCC QC795.34 (ebook) | LCC QC795.34 .R34 2019 (print) | DDC 539.7/520712--dc23
LC record available at *https://lccn.loc.gov/2018046668*

The *Next Generation Science Standards* ("NGSS") were developed by twenty-six states, in collaboration with the National Research Council, the National Science Teachers Association and the American Association for the Advancement of Science in a process managed by Achieve, Inc. For more information go to *www.nextgenscience.org.*

CONTENTS

Part 1: The STEM Road Map: Background, Theory, and Practice

Part 2: Radioactivity: STEM Road Map Module

CONTENTS

ABOUT THE EDITORS AND AUTHORS

Dr. Carla C. Johnson is executive director of the William and Ida Friday Institute for Educational Innovation, associate dean, and professor of science education in the College of Education at North Carolina State University in Raleigh. She was most recently an associate dean, provost fellow, and professor of science education at Purdue University in West Lafayette, Indiana. Dr. Johnson serves as the director of research and evaluation for the Department of Defense–funded Army Educational Outreach Program (AEOP), a global portfolio of STEM education programs, competitions, and apprenticeships. She has been a leader in STEM education for the past decade, serving as the director of STEM Centers, editor of the *School Science and Mathematics* journal, and lead researcher for the evaluation of Tennessee's Race to the Top–funded STEM portfolio. Dr. Johnson has published over 100 articles, books, book chapters, and curriculum books focused on STEM education. She is a former science and social studies teacher and was the recipient of the 2013 Outstanding Science Teacher Educator of the Year award from the Association for Science Teacher Education (ASTE), the 2012 Award for Excellence in Integrating Science and Mathematics from the School Science and Mathematics Association (SSMA), the 2014 award for best paper on Implications of Research for Educational Practice from ASTE, and the 2006 Outstanding Early Career Scholar Award from SSMA. Her research focuses on STEM education policy implementation, effective science teaching, and integrated STEM approaches.

Dr. Janet B. Walton is a research assistant professor and the assistant director of evaluation for AEOP at Purdue University's College of Education. Formerly the STEM workforce program manager for Virginia's Region 2000 and founding director of the Future Focus Foundation, a nonprofit organization dedicated to enhancing the quality of STEM education in the region, she merges her economic development and education backgrounds to develop K–12 curricular materials that integrate real-life issues with sound cross-curricular content. Her research focuses on collaboration between schools and community stakeholders for STEM education and problem- and project-based learning pedagogies. With this research agenda, she works to forge productive relationships between K–12 schools and local business and community stakeholders to bring contextual STEM experiences into the classroom and provide students and educators with innovative resources and curricular materials.

Dr. Erin Peters-Burton is the Donna R. and David E. Sterling endowed professor in science education at George Mason University in Fairfax, Virginia. She uses her experiences from 15 years as an engineer and secondary science, engineering, and mathematics teacher to develop research projects that directly inform classroom practice in science and engineering. Her research agenda is based on the idea that all students should build self-awareness of how they learn science and engineering. She works to help students see themselves as "science-minded" and help teachers create classrooms that support student skills to develop scientific knowledge. To accomplish this, she pursues research projects that investigate ways that students and teachers can use self-regulated learning theory in science and engineering, as well as how inclusive STEM schools can help students succeed. During her tenure as a secondary teacher, she had a National Board Certification in Early Adolescent Science and was an Albert Einstein Distinguished Educator Fellow for NASA. As a researcher, Dr. Peters-Burton has published over 100 articles, books, book chapters, and curriculum books focused on STEM education and educational psychology. She received the Outstanding Science Teacher Educator of the Year award from ASTE in 2016 and a Teacher of Distinction Award and a Scholarly Achievement Award from George Mason University in 2012, and in 2010 she was named University Science Educator of the Year by the Virginia Association of Science Teachers.

Dr. Jennifer Drake-Patrick is an assistant professor of literacy education in the College of Education and Human Development at George Mason University. A former English language arts teacher, she focuses her research on disciplinary literacy.

Dr. Tamara J. Moore is an associate professor of engineering education in the College of Engineering at Purdue University. Dr. Moore's research focuses on defining STEM integration through the use of engineering as the connection and investigating its power for student learning.

Dr. Anthony Pellegrino is an assistant professor of social science in the College of Education at the University of Tennessee, Knoxville. He is a former social studies and history teacher whose research interests include youth-centered pedagogies and social science teacher preparation.

Dr. Bradley D. Rankin is a high school mathematics teacher at Wakefield High School in Arlington, Virginia. He has been teaching mathematics for 20 years, is board certified, and has a PhD in mathematics education leadership from George Mason University.

Dr. Toni A. Sondergeld is an associate professor of assessment, research, and statistics in the School of Education at Drexel University in Philadelphia. Dr. Sondergeld's research concentrates on assessment and evaluation in education, with a focus on K–12 STEM.

ACKNOWLEDGMENTS

This module was developed as a part of the STEM Road Map project (Carla C. Johnson, principal investigator). The Purdue University College of Education, General Motors, and other sources provided funding for this project.

See *www.routledge.com/products/9781138804234* for more information about *STEM Road Map: A Framework for Integrated STEM Education.*

PART 1

THE STEM ROAD MAP

BACKGROUND, THEORY, AND PRACTICE

OVERVIEW OF THE *STEM ROAD MAP CURRICULUM SERIES*

Carla C. Johnson, Erin Peters-Burton, and Tamara J. Moore

The *STEM Road Map Curriculum Series* was conceptualized and developed by a team of STEM educators from across the United States in response to a growing need to infuse real-world learning contexts, delivered through authentic problem-solving pedagogy, into K–12 classrooms. The curriculum series is grounded in integrated STEM, which focuses on the integration of the STEM disciplines—science, technology, engineering, and mathematics—delivered across content areas, incorporating the Framework for 21st Century Learning along with grade-level-appropriate academic standards.

The curriculum series begins in kindergarten, with a five-week instructional sequence that introduces students to the STEM themes and gives them grade-level-appropriate topics and real-world challenges or problems to solve. The series uses project-based and problem-based learning, presenting students with the problem or challenge during the first lesson, and then teaching them science, social studies, English language arts, mathematics, and other content, as they apply what they learn to the challenge or problem at hand.

Authentic assessment and differentiation are embedded throughout the modules. Each *STEM Road Map Curriculum Series* module has a lead discipline, which may be science, social studies, English language arts, or mathematics. All disciplines are integrated into each module, along with ties to engineering. Another key component is the use of STEM Research Notebooks to allow students to track their own learning progress. The modules are designed with a scaffolded approach, with increasingly complex concepts and skills introduced as students progress through grade levels.

The developers of this work view the curriculum as a resource that is intended to be used either as a whole or in part to meet the needs of districts, schools, and teachers who are implementing an integrated STEM approach. A variety of implementation formats are possible, from using one stand-alone module at a given grade level to using all five modules to provide 25 weeks of instruction. Also, within each grade band (K–2, 3–5, 6–8, 9–12), the modules can be sequenced in various ways to suit specific needs.

STANDARDS-BASED APPROACH

The *STEM Road Map Curriculum Series* is anchored in the *Next Generation Science Standards (NGSS)*, the *Common Core State Standards for Mathematics (CCSS Mathematics)*, the *Common Core State Standards for English Language Arts (CCSS ELA)*, and the Framework for 21st Century Learning. Each module includes a detailed curriculum map that incorporates the associated standards from the particular area correlated to lesson plans. The STEM Road Map has very clear and strong connections to these academic standards, and each of the grade-level topics was derived from the mapping of the standards to ensure alignment among topics, challenges or problems, and the required academic standards for students. Therefore, the curriculum series takes a standards-based approach and is designed to provide authentic contexts for application of required knowledge and skills.

THEMES IN THE *STEM ROAD MAP CURRICULUM SERIES*

The K–12 STEM Road Map is organized around five real-world STEM themes that were generated through an examination of the big ideas and challenges for society included in STEM standards and those that are persistent dilemmas for current and future generations:

- Cause and Effect

- Innovation and Progress

- The Represented World

- Sustainable Systems

- Optimizing the Human Experience

These themes are designed as springboards for launching students into an exploration of real-world learning situated within big ideas. Most important, the five STEM Road Map themes serve as a framework for scaffolding STEM learning across the K–12 continuum.

The themes are distributed across the STEM disciplines so that they represent the big ideas in science (Cause and Effect; Sustainable Systems), technology (Innovation and Progress; Optimizing the Human Experience), engineering (Innovation and Progress; Sustainable Systems; Optimizing the Human Experience), and mathematics (The Represented World), as well as concepts and challenges in social studies and 21st century skills that are also excellent contexts for learning in English language arts. The process of developing themes began with the clustering of the *NGSS* performance expectations and the National Academy of Engineering's grand challenges for engineering, which led to the development of the challenge in each module and connections of the module activities to the *CCSS Mathematics* and *CCSS ELA* standards. We performed these mapping processes with large teams of experts and found that these five themes provided breadth, depth, and coherence to frame a high-quality STEM learning experience from kindergarten through 12th grade.

Cause and Effect

The concept of cause and effect is a powerful and pervasive notion in the STEM fields. It is the foundation of understanding how and why things happen as they do. Humans spend considerable effort and resources trying to understand the causes and effects of natural and designed phenomena to gain better control over events and the environment and to be prepared to react appropriately. Equipped with the knowledge of a specific cause-and-effect relationship, we can lead better lives or contribute to the community by altering the cause, leading to a different effect. For example, if a person recognizes that irresponsible energy consumption leads to global climate change, that person can act to remedy his or her contribution to the situation. Although cause and effect is a core idea in the STEM fields, it can actually be difficult to determine. Students should be capable of understanding not only when evidence points to cause and effect but also when evidence points to relationships but not direct causality. The major goal of education is to foster students to be empowered, analytic thinkers, capable of thinking through complex processes to make important decisions. Understanding causality, as well as when it cannot be determined, will help students become better consumers, global citizens, and community members.

Innovation and Progress

One of the most important factors in determining whether humans will have a positive future is innovation. Innovation is the driving force behind progress, which helps create possibilities that did not exist before. Innovation and progress are creative entities, but in the STEM fields, they are anchored by evidence and logic, and they use established concepts to move the STEM fields forward. In creating something new, students must consider what is already known in the STEM fields and apply this knowledge appropriately. When we innovate, we create value that was not there previously and create new conditions and possibilities for even more innovations. Students should consider how their innovations might affect progress and use their STEM thinking to change current human burdens to benefits. For example, if we develop more efficient cars that use by-products from another manufacturing industry, such as food processing, then we have used waste productively and reduced the need for the waste to be hauled away, an indirect benefit of the innovation.

The Represented World

When we communicate about the world we live in, how the world works, and how we can meet the needs of humans, sometimes we can use the actual phenomena to explain a concept. Sometimes, however, the concept is too big, too slow, too small, too fast, or too complex for us to explain using the actual phenomena, and we must use a representation or a model to help communicate the important features. We need representations and models such as graphs, tables, mathematical expressions, and diagrams because it makes our thinking visible. For example, when examining geologic time, we cannot actually observe the passage of such large chunks of time, so we create a timeline or a model that uses a proportional scale to visually illustrate how much time has passed for different eras. Another example may be something too complex for students at a particular grade level, such as explaining the *p* subshell orbitals of electrons to fifth graders. Instead, we use the Bohr model, which more closely represents the orbiting of planets and is accessible to fifth graders.

When we create models, they are helpful because they point out the most important features of a phenomenon. We also create representations of the world with mathematical functions, which help us change parameters to suit the situation. Creating representations of a phenomenon engages students because they are able to identify the important features of that phenomenon and communicate them directly. But because models are estimates of a phenomenon, they leave out some of the details, so it is important for students to evaluate their usefulness as well as their shortcomings.

Sustainable Systems

From an engineering perspective, the term *system* refers to the use of "concepts of component need, component interaction, systems interaction, and feedback. The interaction of subcomponents to produce a functional system is a common lens used by all engineering disciplines for understanding, analysis, and design." (Koehler, Bloom, and Binns 2013, p. 8). Systems can be either open (e.g., an ecosystem) or closed (e.g., a car battery). Ideally, a system should be sustainable, able to maintain equilibrium without much energy from outside the structure. Looking at a garden, we see flowers blooming, weeds sprouting, insects buzzing, and various forms of life living within its boundaries. This is an example of an ecosystem, a collection of living organisms that survive together, functioning as a system. The interaction of the organisms within the system and the influences of the environment (e.g., water, sunlight) can maintain the system for a period of time, thus demonstrating its ability to endure. Sustainability is a desirable feature of a system because it allows for existence of the entity in the long term.

In the STEM Road Map project, we identified different standards that we consider to be oriented toward systems that students should know and understand in the K–12 setting. These include ecosystems, the rock cycle, Earth processes (such as erosion, tectonics, ocean currents, weather phenomena), Earth-Sun-Moon cycles, heat transfer, and the interaction among the geosphere, biosphere, hydrosphere, and atmosphere. Students and teachers should understand that we live in a world of systems that are not independent of each other, but rather are intrinsically linked such that a disruption in one part of a system will have reverberating effects on other parts of the system.

Optimizing the Human Experience

Science, technology, engineering, and mathematics as disciplines have the capacity to continuously improve the ways humans live, interact, and find meaning in the world, thus working to optimize the human experience. This idea has two components: being more suited to our environment and being more fully human. For example, the progression of STEM ideas can help humans create solutions to complex problems, such as improving ways to access water sources, designing energy sources with minimal impact on our environment, developing new ways of communication and expression, and building efficient shelters. STEM ideas can also provide access to the secrets and wonders of nature. Learning in STEM requires students to think logically and systematically, which is a way of knowing the world that is markedly different from knowing the world as an artist. When students can employ various ways of knowing and understand when it is appropriate to use a different way of knowing or integrate ways of knowing, they are fully experiencing the best of what it is to be human. The problem-based learning scenarios provided in the STEM Road Map help students develop ways of thinking like STEM professionals as they ask questions and design solutions. They learn to optimize the human experience by innovating improvements in the designed world in which they live.

THE NEED FOR AN INTEGRATED STEM APPROACH

At a basic level, STEM stands for science, technology, engineering, and mathematics. Over the past decade, however, STEM has evolved to have a much broader scope and broader implications. Now, educators and policy makers refer to STEM as not only a concentrated area for investing in the future of the United States and other nations but also as a domain and mechanism for educational reform.

The good intentions of the recent decade-plus of focus on accountability and increased testing has resulted in significant decreases not only in instructional time for teaching science and social studies but also in the flexibility of teachers to promote authentic, problem solving–focused classroom environments. The shift has had a detrimental impact on student acquisition of vitally important skills, which many refer to as 21st century skills, and often the ability of students to "think." Further, schooling has become increasingly siloed into compartments of mathematics, science, English language arts, and social studies, lacking any of the connections that are overwhelmingly present in

the real world around children. Students have experienced school as content provided in boxes that must be memorized, devoid of any real-world context, and often have little understanding of why they are learning these things.

STEM-focused projects, curriculum, activities, and schools have emerged as a means to address these challenges. However, most of these efforts have continued to focus on the individual STEM disciplines (predominantly science and engineering) through more STEM classes and after-school programs in a "STEM enhanced" approach (Breiner et al. 2012). But in traditional and STEM enhanced approaches, there is little to no focus on other disciplines that are integral to the context of STEM in the real world. Integrated STEM education, on the other hand, infuses the learning of important STEM content and concepts with a much-needed emphasis on 21st century skills and a problem- and project-based pedagogy that more closely mirrors the real-world setting for society's challenges. It incorporates social studies, English language arts, and the arts as pivotal and necessary (Johnson 2013; Rennie, Venville, and Wallace 2012; Roehrig et al. 2012).

FRAMEWORK FOR STEM INTEGRATION IN THE CLASSROOM

The *STEM Road Map Curriculum Series* is grounded in the Framework for STEM Integration in the Classroom as conceptualized by Moore, Guzey, and Brown (2014) and Moore et al. (2014). The framework has six elements, described in the context of how they are used in the *STEM Road Map Curriculum Series* as follows:

1. The STEM Road Map contexts are meaningful to students and provide motivation to engage with the content. Together, these allow students to have different ways to enter into the challenge.

2. The STEM Road Map modules include engineering design that allows students to design technologies (i.e., products that are part of the designed world) for a compelling purpose.

3. The STEM Road Map modules provide students with the opportunities to learn from failure and redesign based on the lessons learned.

4. The STEM Road Map modules include standards-based disciplinary content as the learning objectives.

5. The STEM Road Map modules include student-centered pedagogies that allow students to grapple with the content, tie their ideas to the context, and learn to think for themselves as they deepen their conceptual knowledge.

6. The STEM Road Map modules emphasize 21st century skills and, in particular, highlight communication and teamwork.

All of the STEM Road Map modules incorporate these six elements; however, the level of emphasis on each of these elements varies based on the challenge or problem in each module.

THE NEED FOR THE *STEM ROAD MAP CURRICULUM SERIES*

As focus is increasing on integrated STEM, and additional schools and programs decide to move their curriculum and instruction in this direction, there is a need for high-quality, research-based curriculum designed with integrated STEM at the core. Several good resources are available to help teachers infuse engineering or more STEM enhanced approaches, but no curriculum exists that spans K–12 with an integrated STEM focus. The next chapter provides detailed information about the specific pedagogy, instructional strategies, and learning theory on which the *STEM Road Map Curriculum Series* is grounded.

REFERENCES

Breiner, J., M. Harkness, C. C. Johnson, and C. Koehler. 2012. What is STEM? A discussion about conceptions of STEM in education and partnerships. *School Science and Mathematics* 112 (1): 3–11.

Johnson, C. C. 2013. Conceptualizing integrated STEM education: Editorial. *School Science and Mathematics* 113 (8): 367–368.

Koehler, C. M., M. A. Bloom, and I. C. Binns. 2013. Lights, camera, action: Developing a methodology to document mainstream films' portrayal of nature of science and scientific inquiry. *Electronic Journal of Science Education* 17 (2).

Moore, T. J., S. S. Guzey, and A. Brown. 2014. Greenhouse design to increase habitable land: An engineering unit. *Science Scope* 37 (7): 51–57.

Moore, T. J., M. S. Stohlmann, H.-H. Wang, K. M. Tank, A. W. Glancy, and G. H. Roehrig. 2014. Implementation and integration of engineering in K–12 STEM education. In *Engineering in pre-college settings: Synthesizing research, policy, and practices*, ed. S. Purzer, J. Strobel, and M. Cardella, 35–60. West Lafayette, IN: Purdue Press.

Rennie, L., G. Venville, and J. Wallace. 2012. *Integrating science, technology, engineering, and mathematics: Issues, reflections, and ways forward*. New York: Routledge.

Roehrig, G. H., T. J. Moore, H. H. Wang, and M. S. Park. 2012. Is adding the E enough? Investigating the impact of K–12 engineering standards on the implementation of STEM integration. *School Science and Mathematics* 112 (1): 31–44.

STRATEGIES USED IN THE *STEM ROAD MAP CURRICULUM SERIES*

Erin Peters-Burton, Carla C. Johnson, Toni A. Sondergeld, and Tamara J. Moore

The *STEM Road Map Curriculum Series* uses what has been identified through research as best-practice pedagogy, including embedded formative assessment strategies throughout each module. This chapter briefly describes the key strategies that are employed in the series.

PROJECT- AND PROBLEM-BASED LEARNING

Each module in the *STEM Road Map Curriculum Series* uses either project-based learning or problem-based learning to drive the instruction. Project-based learning begins with a driving question to guide student teams in addressing a contextualized local or community problem or issue. The outcome of project-based instruction is a product that is conceptualized, designed, and tested through a series of scaffolded learning experiences (Blumenfeld et al. 1991; Krajcik and Blumenfeld 2006). Problem-based learning is often grounded in a fictitious scenario, challenge, or problem (Barell 2006; Lambros 2004). On the first day of instruction within the unit, student teams are provided with the context of the problem. Teams work through a series of activities and use open-ended research to develop their potential solution to the problem or challenge, which need not be a tangible product (Johnson 2003).

ENGINEERING DESIGN PROCESS

The *STEM Road Map Curriculum Series* uses engineering design as a way to facilitate integrated STEM within the modules. The engineering design process (EDP) is depicted in Figure 2.1 (p. 10). It highlights two major aspects of engineering design—problem scoping and solution generation—and six specific components of working toward a design: define the problem, learn about the problem, plan a solution, try the solution, test the solution, decide whether the solution is good enough. It also shows that communication

Figure 2.1. Engineering Design Process

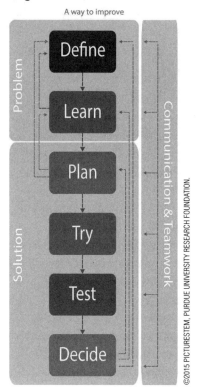

and teamwork are involved throughout the entire process. As the arrows in the figure indicate, the order in which the components of engineering design are addressed depends on what becomes needed as designers progress through the EDP. Designers must communicate and work in teams throughout the process. The EDP is iterative, meaning that components of the process can be repeated as needed until the design is good enough to present to the client as a potential solution to the problem.

Problem scoping is the process of gathering and analyzing information to deeply understand the engineering design problem. It includes defining the problem and learning about the problem. Defining the problem includes identifying the problem, the client, and the end user of the design. The client is the person (or people) who hired the designers to do the work, and the end user is the person (or people) who will use the final design. The designers must also identify the criteria and the constraints of the problem. The criteria are the things the client wants from the solution, and the constraints are the things that limit the possible solutions. The designers must spend significant time learning about the problem, which can include activities such as the following:

- Reading informational texts and researching about relevant concepts or contexts

- Identifying and learning about needed mathematical and scientific skills, knowledge, and tools

- Learning about things done previously to solve similar problems

- Experimenting with possible materials that could be used in the design

Problem scoping also allows designers to consider how to measure the success of the design in addressing specific criteria and staying within the constraints over multiple iterations of solution generation.

Solution generation includes planning a solution, trying the solution, testing the solution, and deciding whether the solution is good enough. Planning the solution includes generating many design ideas that both address the criteria and meet the constraints. Here the designers must consider what was learned about the problem during problem scoping. Design plans include clear communication of design ideas through media such as notebooks, blueprints, schematics, or storyboards. They also include details about the design, such as measurements, materials, colors, costs of materials, instructions for how things fit together, and sets of directions. Making the decision about which design idea to move forward involves considering the trade-offs of each design idea.

Once a clear design plan is in place, the designers must try the solution. Trying the solution includes developing a prototype (a testable model) based on the plan generated. The prototype might be something physical or a process to accomplish a goal. This component of design requires that the designers consider the risk involved in implementing the design. The prototype developed must be tested. Testing the solution includes conducting fair tests that verify whether the plan is a solution that is good enough to meet the client and end user needs and wants. Data need to be collected about the results of the tests of the prototype, and these data should be used to make evidence-based decisions regarding the design choices made in the plan. Here, the designers must again consider the criteria and constraints for the problem.

Using the data gathered from the testing, the designers must decide whether the solution is good enough to meet the client and end user needs and wants by assessment based on the criteria and constraints. Here, the designers must justify or reject design decisions based on the background research gathered while learning about the problem and on the evidence gathered during the testing of the solution. The designers must now decide whether to present the current solution to the client as a possibility or to do more iterations of design on the solution. If they decide that improvements need to be made to the solution, the designers must decide if there is more that needs to be understood about the problem, client, or end user; if another design idea should be tried; or if more planning needs to be conducted on the same design. One way or another, more work needs to be done.

Throughout the process of designing a solution to meet a client's needs and wants, designers work in teams and must communicate to each other, the client, and likely the end user. Teamwork is important in engineering design because multiple perspectives and differing skills and knowledge are valuable when working to solve problems. Communication is key to the success of the designed solution. Designers must communicate their ideas clearly using many different representations, such as text in an engineering notebook, diagrams, flowcharts, technical briefs, or memos to the client.

LEARNING CYCLE

The same format for the learning cycle is used in all grade levels throughout the STEM Road Map, so that students engage in a variety of activities to learn about phenomena in the modules thoroughly and have consistent experiences in the problem- and project-based learning modules. Expectations for learning by younger students are not as high as for older students, but the format of the progression of learning is the same. Students who have learned with curriculum from the STEM Road Map in early grades know

what to expect in later grades. The learning cycle consists of five parts—Introductory Activity/Engagement, Activity/Exploration, Explanation, Elaboration/Application of Knowledge, and Evaluation/Assessment—and is based on the empirically tested 5E model from BSCS (Bybee et al. 2006).

In the Introductory Activity/Engagement phase, teachers introduce the module challenge and use a unique approach designed to pique students' curiosity. This phase gets students to start thinking about what they already know about the topic and begin wondering about key ideas. The Introductory Activity/Engagement phase positions students to be confident about what they are about to learn, because they have prior knowledge, and clues them into what they don't yet know.

In the Activity/Exploration phase, the teacher sets up activities in which students experience a deeper look at the topics that were introduced earlier. Students engage in the activities and generate new questions or consider possibilities using preliminary investigations. Students work independently, in small groups, and in whole-group settings to conduct investigations, resulting in common experiences about the topic and skills involved in the real-world activities. Teachers can assess students' development of concepts and skills based on the common experiences during this phase.

During the Explanation phase, teachers direct students' attention to concepts they need to understand and skills they need to possess to accomplish the challenge. Students participate in activities to demonstrate their knowledge and skills to this point, and teachers can pinpoint gaps in student knowledge during this phase.

In the Elaboration/Application of Knowledge phase, teachers present students with activities that engage in higher-order thinking to create depth and breadth of student knowledge, while connecting ideas across topics within and across STEM. Students apply what they have learned thus far in the module to a new context or elaborate on what they have learned about the topic to a deeper level of detail.

In the last phase, Evaluation/Assessment, teachers give students summative feedback on their knowledge and skills as demonstrated through the challenge. This is not the only point of assessment (as discussed in the section on Embedded Formative Assessments), but it is an assessment of the culmination of the knowledge and skills for the module. Students demonstrate their cognitive growth at this point and reflect on how far they have come since the beginning of the module. The challenges are designed to be multidimensional in the ways students must collaborate and communicate their new knowledge.

STEM RESEARCH NOTEBOOK

One of the main components of the *STEM Road Map Curriculum Series* is the STEM Research Notebook, a place for students to capture their ideas, questions, observations, reflections, evidence of progress, and other items associated with their daily work. At the beginning of each module, the teacher walks students through the setup of the STEM

Research Notebook, which could be a three-ring binder, composition book, or spiral notebook. You may wish to have students create divided sections so that they can easily access work from various disciplines during the module. Electronic notebooks kept on student devices are also acceptable and encouraged. Students will develop their own table of contents and create chapters in the notebook for each module.

Each lesson in the *STEM Road Map Curriculum Series* includes one or more prompts that are designed for inclusion in the STEM Research Notebook and appear as questions or statements that the teacher assigns to students. These prompts require students to apply what they have learned across the lesson to solve the big problem or challenge for that module. Each lesson is designed to meaningfully refer students to the larger problem or challenge they have been assigned to solve with their teams. The STEM Research Notebook is designed to be a key formative assessment tool, as students' daily entries provide evidence of what they are learning. The notebook can be used as a mechanism for dialogue between the teacher and students, as well as for peer and self-evaluation.

The use of the STEM Research Notebook is designed to scaffold student notebooking skills across the grade bands in the *STEM Road Map Curriculum Series*. In the early grades, children learn how to organize their daily work in the notebook as a way to collect their products for future reference. In elementary school, students structure their notebooks to integrate background research along with their daily work and lesson prompts. In the upper grades (middle and high school), students expand their use of research and data gathering through team discussions to more closely mirror the work of STEM experts in the real world.

THE ROLE OF ASSESSMENT IN THE *STEM ROAD MAP CURRICULUM SERIES*

Starting in the middle years and continuing into secondary education, the word *assessment* typically brings grades to mind. These grades may take the form of a letter or a percentage, but they typically are used as a representation of a student's content mastery. If well thought out and implemented, however, classroom assessment can offer teachers, parents, and students valuable information about student learning and misconceptions that does not necessarily come in the form of a grade (Popham 2013).

The *STEM Road Map Curriculum Series* provides a set of assessments for each module. Teachers are encouraged to use assessment information for more than just assigning grades to students. Instead, assessments of activities requiring students to actively engage in their learning, such as student journaling in STEM Research Notebooks, collaborative presentations, and constructing graphic organizers, should be used to move student learning forward. Whereas other curriculum with assessments may include objective-type (multiple-choice or matching) tests, quizzes, or worksheets, we have intentionally avoided these forms of assessments to better align assessment strategies with teacher instruction and

student learning techniques. Since the focus of this book is on project- or problem-based STEM curriculum and instruction that focuses on higher-level thinking skills, appropriate and authentic performance assessments were developed to elicit the most reliable and valid indication of growth in student abilities (Brookhart and Nitko 2008).

Comprehensive Assessment System

Assessment throughout all STEM Road Map curriculum modules acts as a comprehensive system in which formative and summative assessments work together to provide teachers with high-quality information on student learning. Formative assessment occurs when the teacher finds out formally or informally what a student knows about a smaller, defined concept or skill and provides timely feedback to the student about his or her level of proficiency. Summative assessments occur when students have performed all activities in the module and are given a cumulative performance evaluation in which they demonstrate their growth in learning.

A comprehensive assessment system can be thought of as akin to a sporting event. Formative assessments are the practices: It is important to accomplish them consistently, they provide feedback to help students improve their learning, and making mistakes can be worthwhile if students are given an opportunity to learn from them. Summative assessments are the competitions: Students need to be prepared to perform at the best of their ability. Without multiple opportunities to practice skills along the way through formative assessments, students will not have the best chance of demonstrating growth in abilities through summative assessments (Black and Wiliam 1998).

Embedded Formative Assessments

Formative assessments in this module serve two main purposes: to provide feedback to students about their learning and to provide important information for the teacher to inform immediate instructional needs. Providing feedback to students is particularly important when conducting problem- or project-based learning because students take on much of the responsibility for learning, and teachers must facilitate student learning in an informed way. For example, if students are required to conduct research for the Activity/Exploration phase but are not familiar with what constitutes a reliable resource, they may develop misconceptions based on poor information. When a teacher monitors this learning through formative assessments and provides specific feedback related to the instructional goals, students are less likely to develop incomplete or incorrect conceptions in their independent investigations. By using formative assessment to detect problems in student learning and then acting on this information, teachers help move student learning forward through these teachable moments.

Formative assessments come in a variety of formats. They can be informal, such as asking students probing questions related to student knowledge or tasks or simply observing students engaged in an activity to gather information about student skills. Formative assessments can also be formal, such as a written quiz or a laboratory practical. Regardless of the type, three key steps must be completed when using formative assessments (Sondergeld, Bell, and Leusner 2010). First, the assessment is delivered to students so that teachers can collect data. Next, teachers analyze the data (student responses) to determine student strengths and areas that need additional support. Finally, teachers use the results from information collected to modify lessons and create learning environments that reinforce weak points in student learning. If student learning information is not used to modify instruction, the assessment cannot be considered formative in nature.

Formative assessments can be about content, science process skills, or even learning skills. When a formative assessment focuses on content, it assesses student knowledge about the disciplinary core ideas from the *Next Generation Science Standards* (*NGSS*) or content objectives from *Common Core State Standards for Mathematics* (*CCSS Mathematics*) or *Common Core State Standards for English Language Arts* (*CCSS ELA*). Content-focused formative assessments ask students questions about declarative knowledge regarding the concepts they have been learning. Process skills formative assessments examine the extent to which a student can perform science and engineering practices from the *NGSS* or process objectives from *CCSS Mathematics* or *CCSS ELA*, such as constructing an argument. Learning skills can also be assessed formatively by asking students to reflect on the ways they learn best during a module and identify ways they could have learned more.

Assessment Maps

Assessment maps or blueprints can be used to ensure alignment between classroom instruction and assessment. If what students are learning in the classroom is not the same as the content on which they are assessed, the resultant judgment made on student learning will be invalid (Brookhart and Nitko 2008). Therefore, the issue of instruction and assessment alignment is critical. The assessment map for this book (found in Chapter 3) indicates by lesson whether the assessment should be completed as a group or on an individual basis, identifies the assessment as formative or summative in nature, and aligns the assessment with its corresponding learning objectives.

Note that the module includes far more formative assessments than summative assessments. This is done intentionally to provide students with multiple opportunities to practice their learning of new skills before completing a summative assessment. Note also that formative assessments are used to collect information on only one or two learning objectives at a time so that potential relearning or instructional modifications can focus on smaller and more manageable chunks of information. Conversely, summative assessments in the module cover many more learning objectives, as they are traditionally used as final markers of student learning. This is not to say that information collected from summative assessments cannot or should not be used formatively. If teachers find that gaps in student

learning persist after a summative assessment is completed, it is important to revisit these existing misconceptions or areas of weakness before moving on (Black et al. 2003).

SELF-REGULATED LEARNING THEORY IN THE STEM ROAD MAP MODULES

Many learning theories are compatible with the STEM Road Map modules, such as constructivism, situated cognition, and meaningful learning. However, we feel that the self-regulated learning theory (SRL) aligns most appropriately (Zimmerman 2000). SRL requires students to understand that thinking needs to be motivated and managed (Ritchhart, Church, and Morrison 2011). The STEM Road Map modules are student centered and are designed to provide students with choices, concrete hands-on experiences, and opportunities to see and make connections, especially across subjects (Eliason and Jenkins 2012; NAEYC 2016). Additionally, SRL is compatible with the modules because it fosters a learning environment that supports students' motivation, enables students to become aware of their own learning strategies, and requires reflection on learning while experiencing the module (Peters and Kitsantas 2010).

The theory behind SRL (see Figure 2.2) explains the different processes that students engage in before, during, and after a learning task. Because SRL is a cyclical learning process, the accomplishment of one cycle develops strategies for the next learning cycle. This cyclic way of learning aligns with the various sections in the STEM Road Map lesson plans on Introductory Activity/ Engagement, Activity/Exploration, Explanation, Elaboration/Application of Knowledge, and Evaluation/Assessment. Since the students engaged in a module take on much of the responsibility for learning, this theory also provides guidance for teachers to keep students on the right track.

The remainder of this section explains how SRL theory is embedded within the five sections of each module and points out ways to

Figure 2.2. SRL Theory

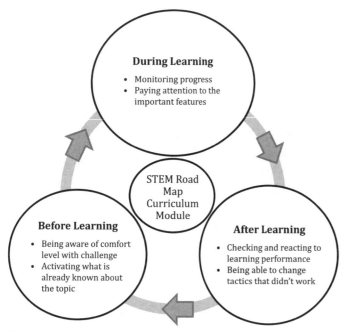

Source: Adapted from Zimmerman 2000.

support students in becoming independent learners of STEM while productively functioning in collaborative teams.

Before Learning: Setting the Stage

Before attempting a learning task such as the STEM Road Map modules, teachers should develop an understanding of their students' level of comfort with the process of accomplishing the learning and determine what they already know about the topic. When students are comfortable with attempting a learning task, they tend to take more risks in learning and as a result achieve deeper learning (Bandura 1986).

The STEM Road Map curriculum modules are designed to foster excitement from the very beginning. Each module has an Introductory Activity/Engagement section that introduces the overall topic from a unique and exciting perspective, engaging the students to learn more so that they can accomplish the challenge. The Introductory Activity also has a design component that helps teachers assess what students already know about the topic of the module. In addition to the deliberate designs in the lesson plans to support SRL, teachers can support a high level of student comfort with the learning challenge by finding out if students have ever accomplished the same kind of task and, if so, asking them to share what worked well for them.

During Learning: Staying the Course

Some students fear inquiry learning because they aren't sure what to do to be successful (Peters 2010). However, the STEM Road Map curriculum modules are embedded with tools to help students pay attention to knowledge and skills that are important for the learning task and to check student understanding along the way. One of the most important processes for learning is the ability for learners to monitor their own progress while performing a learning task (Peters 2012). The modules allow students to monitor their progress with tools such as the STEM Research Notebooks, in which they record what they know and can check whether they have acquired a complete set of knowledge and skills. The STEM Road Map modules support inquiry strategies that include previewing, questioning, predicting, clarifying, observing, discussing, and journaling (Morrison and Milner 2014). Through the use of technology throughout the modules, inquiry is supported by providing students access to resources and data while enabling them to process information, report the findings, collaborate, and develop 21st century skills.

It is important for teachers to encourage students to have an open mind about alternative solutions and procedures (Milner and Sondergeld 2015) when working through the STEM Road Map curriculum modules. Novice learners can have difficulty knowing what to pay attention to and tend to treat each possible avenue for information as equal (Benner 1984). Teachers are the mentors in a classroom and can point out ways for students to approach learning during the Activity/Exploration, Explanation, and

Elaboration/Application of Knowledge portions of the lesson plans to ensure that students pay attention to the important concepts and skills throughout the module. For example, if a student is to demonstrate conceptual awareness of motion when working on roller coaster research, but the student has misconceptions about motion, the teacher can step in and redirect student learning.

After Learning: Knowing What Works

The classroom is a busy place, and it may often seem that there is no time for self-reflection on learning. Although skipping this reflective process may save time in the short term, it reduces the ability to take into account things that worked well and things that didn't so that teaching the module may be improved next time. In the long run, SRL skills are critical for students to become independent learners who can adapt to new situations. By investing the time it takes to teach students SRL skills, teachers can save time later, because students will be able to apply methods and approaches for learning that they have found effective to new situations. In the Evaluation/Assessment portion of the STEM Road Map curriculum modules, as well as in the formative assessments throughout the modules, two processes in the after-learning phase are supported: evaluating one's own performance and accounting for ways to adapt tactics that didn't work well. Students have many opportunities to self-assess in formative assessments, both in groups and individually, using the rubrics provided in the modules.

The designs of the *NGSS* and *CCSS* allow for students to learn in diverse ways, and the STEM Road Map curriculum modules emphasize that students can use a variety of tactics to complete the learning process. For example, students can use STEM Research Notebooks to record what they have learned during the various research activities. Notebook entries might include putting objectives in students' own words, compiling their prior learning on the topic, documenting new learning, providing proof of what they learned, and reflecting on what they felt successful doing and what they felt they still needed to work on. Perhaps students didn't realize that they were supposed to connect what they already knew with what they learned. They could record this and would be prepared in the next learning task to begin connecting prior learning with new learning.

SAFETY IN STEM

Student safety is a primary consideration in all subjects but is an area of particular concern in science, where students may interact with unfamiliar tools and materials that may pose additional safety risks. It is important to implement safety practices within the context of STEM investigations, whether in a classroom laboratory or in the field. When you keep safety in mind as a teacher, you avoid many potential issues with the lesson while also protecting your students.

STEM safety practices encompass things considered in the typical science classroom. Ensure that students are familiar with basic safety considerations, such as wearing protective equipment (e.g., safety glasses or goggles and latex-free gloves) and taking care with sharp objects, and know emergency exit procedures. Teachers should learn beforehand the locations of the safety eyewash, fume hood, fire extinguishers, and emergency shut-off switch in the classroom and how to use them. Also be aware of any school or district safety policies that are in place and apply those that align with the work being conducted in the lesson. It is important to review all safety procedures annually.

STEM investigations should always be supervised. Each lesson in the modules includes teacher guidelines for applicable safety procedures that should be followed. Before each investigation, teachers should go over these safety procedures with the student teams. Some STEM focus areas such as engineering require that students can demonstrate how to properly use equipment in the maker space before the teacher allows them to proceed with the lesson.

Information about classroom science safety, including a safety checklist for science classrooms, general lab safety recommendations, and links to other science safety resources, is available at the Council of State Science Supervisors (CSSS) website at *www.csss-science. org/safety.shtml.* The National Science Teachers Association (NSTA) provides a list of science rules and regulations, including standard operating procedures for lab safety, and a safety acknowledgment form for students and parents or guardians to sign. You can access these resources at *http://static.nsta.org/pdfs/SafetyInTheScienceClassroom.pdf.* In addition, NSTA's Safety in the Science Classroom web page (*www.nsta.org/safety*) has numerous links to safety resources, including papers written by the NSTA Safety Advisory Board.

Disclaimer: The safety precautions for each activity are based on use of the recommended materials and instructions, legal safety standards, and better professional practices. Using alternative materials or procedures for these activities may jeopardize the level of safety and therefore is at the user's own risk.

REFERENCES

Bandura, A. 1986. *Social foundations of thought and action: A social cognitive theory.* Englewood Cliffs, NJ: Prentice-Hall.

Barell, J. 2006. *Problem-based learning: An inquiry approach.* Thousand Oaks, CA: Corwin Press.

Benner, P. 1984. *From novice to expert: Excellence and power in clinical nursing practice.* Menlo Park, CA: Addison-Wesley.

Black, P., C. Harrison, C. Lee, B. Marshall, and D. Wiliam. 2003. *Assessment for learning: Putting it into practice.* Berkshire, UK: Open University Press.

Black, P., and D. Wiliam. 1998. Inside the black box: Raising standards through classroom assessment. *Phi Delta Kappan* 80 (2): 139–148.

Blumenfeld, P., E. Soloway, R. Marx, J. Krajcik, M. Guzdial, and A. Palincsar. 1991. Motivating project-based learning: Sustaining the doing, supporting learning. *Educational Psychologist* 26 (3): 369–398.

Brookhart, S. M., and A. J. Nitko. 2008. *Assessment and grading in classrooms.* Upper Saddle River, NJ: Pearson.

Bybee, R., J. Taylor, A. Gardner, P. Van Scotter, J. Carlson Powell, A. Westbrook, and N. Landes. 2006. *The BSCS 5E instructional model: Origins and effectiveness.* Colorado Springs, CO: BSCS.

Eliason, C. F., and L. T. Jenkins. 2012. *A practical guide to early childhood curriculum.* 9th ed. New York: Merrill.

Johnson, C. 2003. Bioterrorism is real-world science: Inquiry-based simulation mirrors real life. *Science Scope* 27 (3): 19–23.

Krajcik, J., and P. Blumenfeld. 2006. Project-based learning. In *The Cambridge handbook of the learning sciences,* ed. R. Keith Sawyer, 317–334. New York: Cambridge University Press.

Lambros, A. 2004. *Problem-based learning in middle and high school classrooms: A teacher's guide to implementation.* Thousand Oaks, CA: Corwin Press.

Milner, A. R., and T. Sondergeld. 2015. Gifted urban middle school students: The inquiry continuum and the nature of science. *National Journal of Urban Education and Practice* 8 (3): 442–461.

Morrison, V., and A. R. Milner. 2014. Literacy in support of science: A closer look at cross-curricular instructional practice. *Michigan Reading Journal* 46 (2): 42–56.

National Association for the Education of Young Children (NAEYC). 2016. Developmentally appropriate practice position statements. *www.naeyc.org/positionstatements/dap.*

Peters, E. E. 2010. Shifting to a student-centered science classroom: An exploration of teacher and student changes in perceptions and practices. *Journal of Science Teacher Education* 21 (3): 329–349.

Peters, E. E. 2012. Developing content knowledge in students through explicit teaching of the nature of science: Influences of goal setting and self-monitoring. *Science and Education* 21 (6): 881–898.

Peters, E. E., and A. Kitsantas. 2010. The effect of nature of science metacognitive prompts on science students' content and nature of science knowledge, metacognition, and self-regulatory efficacy. *School Science and Mathematics* 110: 382–396.

Popham, W. J. 2013. *Classroom assessment: What teachers need to know.* 7th ed. Upper Saddle River, NJ: Pearson.

Ritchhart, R., M. Church, and K. Morrison. 2011. *Making thinking visible: How to promote engagement, understanding, and independence for all learners.* San Francisco, CA: Jossey-Bass.

Sondergeld, T. A., C. A. Bell, and D. M. Leusner. 2010. Understanding how teachers engage in formative assessment. *Teaching and Learning* 24 (2): 72–86.

Zimmerman, B. J. 2000. Attaining self-regulation: A social-cognitive perspective. In *Handbook of self-regulation,* ed. M. Boekaerts, P. Pintrich, and M. Zeidner, 13–39. San Diego: Academic Press.

PART 2

RADIOACTIVITY

STEM ROAD MAP MODULE

RADIOACTIVITY MODULE OVERVIEW

Janet B. Walton, Bradley D. Rankin, Erin Peters-Burton, Anthony Pellegrino, Jennifer Drake-Patrick, and Carla C. Johnson

THEME: The Represented World

LEAD DISCIPLINES: Science and Mathematics

MODULE SUMMARY

Since the mid-1940s, interest in harnessing radioactivity as a source of electricity has altered the energy production industry. The understanding that radioactive materials could be used as alternatives to fossil fuels sparked an ongoing societal debate about the safety and efficiency of nuclear energy that continues today. In this module, students consider this debate from various perspectives as they investigate the science, history, and public health implications of nuclear energy. Students learn about nuclear fission and the process of radioactive decay, as well as the history and societal implications of using nuclear energy as a power source, and also have the opportunity to explore nuclear fusion and its potential as an energy source. Through research and inquiry, including accessing primary documents related to the development and use of nuclear technologies and creating models of radioactive decay, nuclear fission, and nuclear fusion, student teams gain an understanding of the potential and dangers of atomic energy. Using this understanding and the engineering design process (EDP), student teams assume the role of various stakeholder groups in creating a response to a fictional nuclear power plant accident (adapted from Peters-Burton et al. 2015).

ESTABLISHED GOALS AND OBJECTIVES

At the conclusion of this module, students will be able to do the following:

- Create models of nuclear fission using physical models, computer-generated simulations, mathematical computations, and other student-created methods

- Calculate the energy yield of an individual nuclear event (decay, fission, and fusion), and use exponential functions to represent chain reactions

- Explain how radioactive decay, nuclear fission, and nuclear fusion work

- Identify the safety and environmental concerns related to using nuclear fission in power-generating plants

- Explain how a pressurized water nuclear reactor works

- Explain the history of the use of nuclear energy and identify key milestones and events that influence societal perspectives on nuclear energy use

- Apply understanding of the science and history of nuclear energy to take a particular stakeholder group stance and create a solution to a problem presented in the form of a fictional nuclear accident

- Use the EDP to formulate a solution to a complex problem

- Collaborate with peers to solve a problem

The lessons in this module take into account that it may not be possible for a teacher to collaborate with teachers from other content areas or that teachers from two different subject areas may not have the same students, so teaching in an integrated way in each class may not make sense. Therefore, the lessons are written so that the science teacher can teach the science classes and only a little of each other content area. That is, if you are teaching the module alone, you may choose to follow only the lead subject, offering enrichment activities in the other connecting subjects. You may want to collaborate with peers in the other subjects to get ideas for ways to incorporate the supporting connections seamlessly. If you are able to teach an integrated curriculum, you can use the module as written for all four subjects in each of the Learning Components sections of the module.

CHALLENGE OR PROBLEM FOR STUDENTS TO SOLVE: GAMMATOWN CRISIS CHALLENGE

Student teams are challenged to take on the roles of stakeholder groups to create responses to a fictional nuclear accident in Gammatown, U.S.A., in the Gammatown Crisis Challenge. Teams synthesize their understanding of the history, science, and current context of nuclear energy use to prepare a presentation for a target audience from the viewpoint of a stakeholder group and to create a model of a scientific phenomenon or technological object or process related to their objective.

Driving Question: How does the use of nuclear energy to meet our energy demands affect society?

CONTENT STANDARDS ADDRESSED IN THIS STEM ROAD MAP MODULE

A full listing with descriptions of the standards this module addresses can be found in the appendix. Listings of the particular standards addressed within lessons are provided in a table for each lesson in Chapter 4.

STEM RESEARCH NOTEBOOK

Each student should maintain a STEM Research Notebook, which will serve as a place for students to organize their work throughout this module (see p. 12 for more general discussion on setup and use of the notebook). All written work in the module should be included in the notebook, including records of students' thoughts and ideas, fictional accounts based on the concepts in the module, and records of student progress through the EDP. The notebooks may be maintained across subject areas, giving students the opportunity to see that although their classes may be separated during the school day, the knowledge they gain is connected. You may also wish to have students include the STEM Research Notebook Guidelines student handout on page 26 in their notebooks.

Emphasize to students the importance of organizing all information in a Research Notebook. Explain to them that scientists and other researchers maintain detailed Research Notebooks in their work. These notebooks, which are crucial to researchers' work because they contain critical information and track the researchers' progress, are often considered legal documents for scientists who are pursuing patents or wish to provide proof of their discovery process.

STUDENT HANDOUT

STEM RESEARCH NOTEBOOK GUIDELINES

STEM professionals record their ideas, inventions, experiments, questions, observations, and other work details in notebooks so that they can use these notebooks to help them think about their projects and the problems they are trying to solve. You will each keep a STEM Research Notebook during this module that is like the notebooks that STEM professionals use. In this notebook, you will include all your work and notes about ideas you have. The notebook will help you connect your daily work with the big problem or challenge you are working to solve.

It is important that you organize your notebook entries under the following headings:

1. **Chapter Topic or Title of Problem or Challenge:** You will start a new chapter in your STEM Research Notebook for each new module. This heading is the topic or title of the big problem or challenge that your team is working to solve in this module.

2. **Date and Topic of Lesson Activity for the Day:** Each day, you will begin your daily entry by writing the date and the day's lesson topic at the top of a new page. Write the page number both on the page and in the table of contents.

3. **Information Gathered From Research:** This is information you find from outside resources such as websites or books.

4. **Information Gained From Class or Discussions With Team Members:** This information includes any notes you take in class and notes about things your team discusses. You can include drawings of your ideas here, too.

5. **New Data Collected From Investigations:** This includes data gathered from experiments, investigations, and activities in class.

6. **Documents:** These are handouts and other resources you may receive in class that will help you solve your big problem or challenge. Paste or staple these documents in your STEM Research Notebook for safekeeping and easy access later.

7. **Personal Reflections:** Here, you record your own thoughts and ideas on what you are learning.

8. **Lesson Prompts:** These are questions or statements that your teacher assigns you within each lesson to help you solve your big problem or challenge. You will respond to the prompts in your notebook.

9. **Other Items:** This section includes any other items your teacher gives you or other ideas or questions you may have.

NATIONAL SCIENCE TEACHERS ASSOCIATION

MODULE LAUNCH

Show students photos or videos depicting various uses of nuclear energy as well as some of the well-publicized nuclear power plant accidents (Three Mile Island, Chernobyl, and Fukushima). Have students share their prior understanding of nuclear energy and discuss their current opinions of nuclear energy as an electric power source and whether they believe it should be used, asking questions such as the following:

- What is nuclear energy?
- Is it safe?
- Is it clean?
- Is it renewable?
- Why was it first used?
- What can it do for humans in the future?

Next, show students a video about nuclear energy such as "The Future of Clean Nuclear Energy Is Coming" at *www.youtube.com/watch?v=t7FvxN_gkt4*. Then, introduce students to the module challenge, the Gammatown Crisis Challenge, and have students work in groups to create lists of knowledge they think they will need to address the challenge.

PREREQUISITE SKILLS FOR THE MODULE

Students enter this module with a wide range of pre-existing skills, information, and knowledge. Table 3.1 (p. 28) provides an overview of prerequisite skills and knowledge that students are expected to apply in this module, along with examples of how they apply this knowledge throughout the module. Differentiation strategies are also provided for students who may need additional support in acquiring or applying this knowledge.

Table 3.1. Prerequisite Key Knowledge and Examples of Applications and Differentiation Strategies

Prerequisite Key Knowledge	Application of Knowledge by Students	Differentiation for Students Needing Knowledge
Science • Understand subatomic particles (electrons, neutrons, protons). • Able to read the periodic table of elements and identify atomic numbers and atomic weights. • Understand that isotopes of a given element vary in their numbers of neutrons and thus their stability.	*Science* • Apply knowledge of elements and isotopes to understand and model nuclear reactions.	*Science* • Provide students with resources about the periodic table and atomic composition (tutorial videos, websites, tables with additional explanations), as well as models of radioactive elements used in nuclear processes. • Conduct a short lesson about the periodic table, reviewing atomic number, atomic weight, and isotopes.
Mathematics • Understand exponential functions. • Able to express quantities using scientific notation.	*Mathematics* • Apply exponential functions to model the decay of radioactive elements and the energy yields in fission and fusion reactions. • Use scientific notation to express the amount of energy released from nuclear reactions.	*Mathematics* • Model the use of exponential growth equations. • Provide problems to work through as a class that use exponential growth and decay, such as compound interest (growth) and the atmospheric pressure at increasingly higher altitudes (decay). • Review procedures for using scientific notation with the class; provide sample problems for students to work through, both as a class and individually; provide online tutorials about scientific notation.

POTENTIAL STEM MISCONCEPTIONS

Students enter the classroom with a wide variety of prior knowledge and ideas, so it is important to be alert to misconceptions, or inappropriate understandings of foundational knowledge. These misconceptions can be classified as one of several types: "preconceived notions," opinions based on popular beliefs or understandings; "nonscientific beliefs," knowledge students have gained about science from sources outside the scientific community; "conceptual misunderstandings," incorrect conceptual models based on incomplete understanding of concepts; "vernacular misconceptions," misunderstandings of words based on their common use versus their scientific use; and "factual misconceptions," incorrect or imprecise knowledge learned in early life that remains unchallenged (NRC 1997, p. 28). Misconceptions must be addressed and dismantled in order for students to reconstruct their knowledge, and therefore teachers should be prepared to take the following steps:

- *Identify students' misconceptions.*

- *Provide a forum for students to confront their misconceptions.*

- *Help students reconstruct and internalize their knowledge, based on scientific models.*
  (NRC 1997, p. 29)

Keeley and Harrington (2010) recommend using diagnostic tools such as probes and formative assessment to identify and confront student misconceptions and begin the process of reconstructing student knowledge. Keeley's *Uncovering Student Ideas in Science* series contains probes targeted toward uncovering student misconceptions in a variety of areas and may be useful resources for addressing student misconceptions in this module.

Some commonly held misconceptions specific to lesson content are provided with each lesson so that you can be alert for student misunderstanding of the science concepts presented and used during this module. The American Association for the Advancement of Science has also identified misconceptions that students frequently hold regarding various science concepts (see the links at *http://assessment.aaas.org/topics*).

SRL PROCESS COMPONENTS

Table 3.2 (p. 30) illustrates some of the activities in the Radioactivity module and how they align with the self-regulated learning (SRL) process before, during, and after learning.

Table 3.2. SRL Process Components

Learning Process Components	Example From Radioactivity Module	Lesson Number and Learning Component
BEFORE LEARNING		
Motivates students	A class discussion about nuclear energy gives students the opportunity to share their current understanding of nuclear energy and discuss their opinions about nuclear energy as a power source.	Lesson 1, Introductory Activity/Engagement
Evokes prior learning	Students connect their knowledge about the Manhattan Project and nuclear weapons development to understand the development of nuclear energy for peaceful (energy production) purposes.	Lesson 2, Activity/ Exploration
DURING LEARNING		
Focuses on important features	Students compare the energy output of a fission reaction with their typical yearly home energy consumption.	Lesson 2, Activity/ Exploration
Helps students monitor their progress	Students respond to a STEM Research Notebook prompt in which they apply their learning about nuclear energy to express and support with evidence an opinion about the safety of nuclear power plants.	Lesson 2, Elaboration/ Application of Knowledge
AFTER LEARNING		
Evaluates learning	Students are challenged to analyze the response to a historic nuclear accident using what they have learned about nuclear science and technology.	Lesson 4, Activity/ Exploration
Takes account of what worked and what did not work	Students practice their presentations for their Gammatown Crisis Challenge response and incorporate feedback to improve their presentations.	Lesson 5, Elaboration/ Application of Knowledge

STRATEGIES FOR DIFFERENTIATING INSTRUCTION WITHIN THIS MODULE

For the purposes of this curriculum module, differentiated instruction is conceptualized as a way to tailor instruction—including process, content, and product—to various student needs in your class. A number of differentiation strategies are integrated into lessons across the module. The problem- and project-based learning approach used in the lessons is designed to address students' multiple intelligences by providing a variety of entry points and methods to investigate the key concepts in the module. Differentiation strategies for students needing support in prerequisite knowledge can be found in Table 3.1 (p. 28). You are encouraged to use information gained about student prior knowledge during introductory activities and discussions to inform your instructional differentiation. Strategies incorporated into this lesson include flexible grouping, varied environmental learning contexts, assessments, compacting, and tiered assignments and scaffolding.

Flexible Grouping. Students work collaboratively in a variety of activities throughout this module. Grouping strategies you might employ include student-led grouping, grouping students according to ability level or common interests, grouping students randomly, or grouping them so that students in each group have complementary strengths (for instance, one student might be strong in mathematics, another in art, and another in writing).

Varied Environmental Learning Contexts. Students have the opportunity to learn in various contexts throughout the module, including alone, in groups, in quiet reading and research-oriented activities, and in active learning through inquiry and design activities. In addition, students learn in a variety of ways, including through doing inquiry activities, journaling, reading texts, watching videos, participating in class discussion, and conducting web-based research.

Assessments. Students are assessed in a variety of ways throughout the module, including individual and collaborative formative and summative assessments. Students have the opportunity to produce work via written text, oral and media presentations, and modeling. You may choose to provide students with additional choices of media for their products (for example, PowerPoint presentations, posters, or student-created websites or blogs).

Compacting. Based on student prior knowledge, you may wish to adjust instructional activities for students who exhibit prior mastery of a learning objective. You may wish to compile a classroom database of research resources and supplementary readings for a variety of reading levels and on a variety of topics related to the module's topic to provide opportunities for students to undertake independent reading.

Tiered Assignments and Scaffolding. Based on your awareness of student ability, understanding of concepts, and mastery of skills, you may wish to provide students with

variations on activities by adding complexity to assignments or providing more or fewer learning supports for activities throughout the module. For instance, some students may need additional support in identifying key search words and phrases for web-based research or may benefit from cloze sentence handouts to enhance vocabulary understanding. Other students may benefit from expanded reading selections and additional reflective writing or from working with manipulatives and other visual representations of mathematical concepts. You may also work with your school librarian to compile a set of topical resources at a variety of reading levels.

STRATEGIES FOR ENGLISH LANGUAGE LEARNERS

Students who are developing proficiency in English language skills require additional supports to simultaneously learn academic content and the specialized language associated with specific content areas. WIDA (2012) has created a framework for providing support to these students and makes available rubrics and guidance on differentiating instructional materials for English language learners (ELLs). In particular, ELL students may benefit from additional sensory supports such as images, physical modeling, and graphic representations of module content, as well as interactive support through collaborative work. This module offers ongoing opportunities for ELL students to work collaboratively. The focus on the production of electrical energy affords an opportunity for ELL students to share culturally diverse experiences with primary energy sources, including fossil fuels, nuclear energy, and renewables such as wind and solar energy.

When differentiating instruction for ELL students, you should carefully consider the needs of these students as you introduce and use academic language in various language domains (listening, speaking, reading, and writing) throughout this module. To adequately differentiate instruction for ELL students, you should have an understanding of the proficiency level of each student. The following 9–12 WIDA standards are relevant to this module:

- Standard 1: Social and Instructional Language. Focus on study skills and strategies, information gathering, workplace readiness.

- Standard 2: The language of Language Arts. Focus on autobiographical and biographical narratives, critical commentary, research, note taking.

- Standard 3: The language of Mathematics. Focus on coordinate planes, graphs, and equations; data displays and interpretation; mathematical relations and functions; problem solving; scale and proportion.

- Standard 4: The language of Science. Focus on atoms and molecules/nuclear structures; chemical and physical change; conservation of energy and matter; elements and compounds; nuclear change; scientific research and investigation.

- Standard 5: The language of Social Studies. Focus on behaviors of individuals and groups; historical figures and times; the story of the United States.

SAFETY CONSIDERATIONS FOR THE ACTIVITIES IN THIS MODULE

For precautions, see the specific safety notes after the list of materials in each lesson. For more general safety guidelines, see the Safety in STEM section in Chapter 2 (p. 18).

DESIRED OUTCOMES AND MONITORING SUCCESS

The desired outcomes for this module are outlined in Table 3.3, along with suggested ways to gather evidence to monitor student success. For more specific details on desired outcomes, see the Established Goals and Objectives sections for the module and individual lessons.

Table 3.3. Desired Outcomes and Evidence of Success in Achieving Identified Outcomes

Desired Outcome	Evidence of Success	
	Performance Tasks	Other Measures
Students can apply an understanding of exponential functions, nuclear reactions, and the societal implications of the use of nuclear reactions to produce electrical energy to respond to a fictional scenario from a stakeholder group perspective.	• Students maintain STEM Research Notebooks that contain data from investigations, sketches, ideas, questions, and research notes. • Students create models of nuclear fission and fusion. • Students create models of scientific phenomena or technological items related to the team challenge. • Students describe the production of electrical energy using nuclear reactions. • Students respond to a fictional nuclear power plant accident from a specific stakeholder group's perspective. • Students are assessed using project rubrics that focus on content and application of skills related to academic content.	Student collaboration is evaluated using a collaboration rubric.

ASSESSMENT PLAN OVERVIEW AND MAP

Table 3.4 provides an overview of the major group and individual *products* and *deliverables*, or things that student teams will produce in this module, that constitute the assessment for this module. See Table 3.5 for a full assessment map of formative and summative assessments in this module.

Table 3.4. Major Products and Deliverables in Lead Disciplines for Groups and Individuals

Lesson	Major Group Products and Deliverables	Major Individual Products and Deliverables
1	• Radioactive Decay Chain Models • Group contribution to Radioactivity Timeline	• STEM Research Notebook entries • Sweetium Half-Life handout and graphs
2	• Nuclear reactor schematic diagram • Nuclear Fission Model • Thorium reactor jigsaw essay • Nuclear history presentations	• STEM Research Notebook entries • Letter to President Roosevelt • Evidence of collaboration
3	• Nuclear Fusion Model • Product Development Challenge product and presentation	• STEM Research Notebook entries • Evidence of collaboration
4	• Anatomy of an Accident team posters and presentations	• STEM Research Notebook entries • Anatomy of an Accident Study Guide • Evidence of collaboration
5	• Gammatown Crisis stakeholder group presentation • Gammatown Crisis stakeholder group printed materials • Gammatown Crisis stakeholder group prototype	• Engineering design process entries in STEM Research Notebook • Evidence of collaboration

Table 3.5. Assessment Map for Radioactivity Module

Lesson	Assessment	Group/ Individual	Formative/ Summative	Lesson Objective Assessed
1	Radioactivity *timeline*	Group	Formative	• Discuss how radioactive decay was discovered and how scientific understanding of the process progressed to our present-day understanding. • Describe how the minuscule amount of energy released in a single alpha, beta, or gamma decay is significant because of the proportion of atoms in matter. • Discuss the major nuclear power plant accidents of the 20th and 21st centuries, and identify implications of these accidents. • Identify several uses of radioactive elements, and imagine ideas for future use.
1	Radioactive Decay Chain Model *rubric*	Group	Formative	• Model the decay chain of a radioactive element. • Apply the engineering design process (EDP) to solve a problem. • Collaborate with peers to solve a problem.
1	STEM Research Notebook *prompt*	Individual	Formative	• Understand that radioactive decay is an exponential function. • Create real-world examples of exponential functions other than radioactive decay.
2	Nuclear Fission Model *rubric*	Group	Summative	• Understand that fission is a nuclear process in which a large atomic nucleus splits into smaller nuclei, releasing the energy stored in the nuclear bonds. • Understand that some fission reactions occur spontaneously, while others, such as those used in nuclear reactors, require an energy input. • Understand that a nuclear chain reaction occurs when neutrons released during fission cause fission in one or more other nuclei. • Create a model of nuclear fission. • Apply the EDP to solve a complex problem. • Collaborate with peers to solve a problem.

Continued

Table 3.5. (*continued*)

Lesson	Assessment	Group/ Individual	Formative/ Summative	Lesson Objective Assessed
2	Thorium Reactor Jigsaw Essay *rubric*	Group	Formative	• Identify thorium as an alternative to uranium for nuclear fission reactions, and discuss the advantages and disadvantages of using thorium in nuclear reactors. • Collaborate with peers to solve a problem.
2	Letter to President Roosevelt STEM Research Notebook *prompt*	Individual	Formative	• Apply understanding of the history of nuclear science to understand the current scientific and environmental context of using nuclear power to meet society's energy needs. • Understand that fission is a nuclear process in which a large atomic nucleus splits into smaller nuclei, releasing the energy stored in the nuclear bonds.
2	Team nuclear history *presentations*	Group	Formative	• Apply understanding of the history of nuclear science to understand the current scientific and environmental context of using nuclear power to meet society's energy needs.
2	STEM Research Notebook *prompt*	Individual	Formative	• Understand that there are ecological implications of using fission reactions to produce electricity.
3	Nuclear Fusion Model *rubric*	Group	Formative	• Explain the differences between fission and fusion reactions. • Explain the potential of nuclear fusion as a power source. • Describe the challenges associated with using nuclear fusion as a power source. • Create a model of nuclear fusion, exhibiting an understanding of why intense heat and pressure are necessary to generate a fusion reaction. • Apply the EDP to solve a complex problem. • Collaborate with peers to solve a problem.
3	Product Development Challenge *rubric*	Group	Summative	• Use their understanding of fusion energy to propose a novel product. • Apply the EDP to solve a complex problem. • Collaborate with peers to solve a problem.

Continued

Table 3.5. (*continued*)

Lesson	Assessment	Group/ Individual	Formative/ Summative	Lesson Objective Assessed
3	STEM Research Notebook *prompt*	Individual	Formative	• Use understanding of fission and fusion to clearly explain to others the product developed by student's team in challenge.
4	Anatomy of an Accident *study guide*	Individual	Formative	• Use primary source documents to understand the Three Mile Island nuclear accident. • Apply understanding of nuclear fission reactions to provide a scientific explanation for the accident. • Analyze the response of various stakeholder groups to the accident.
4	Anatomy of an Accident *posters and presentations*	Group	Formative	• Apply their understanding of the Three Mile Island nuclear accident to create a poster providing an overview of one aspect of the accident.
4	STEM Research Notebook *prompt*	Individual	Formative	• Solve mathematical problems related to the Three Mile Island nuclear accident using dimensional analysis.
5	Stakeholder Group Presentation *rubric*	Group	Summative	• Apply understanding of nuclear energy to create a stakeholder-specific response to a fictional nuclear accident. • Demonstrate understanding of nuclear science in the presentation. • Demonstrate understanding of the mathematics concepts introduced in the module. • Use mathematical modeling to convey information about nuclear reactions. • Use persuasive language to present an argument to a target audience. • Apply the EDP to solve a complex problem. • Collaborate with peers to solve a problem.

Continued

Table 3.5. (*continued*)

Lesson	Assessment	Group/Individual	Formative/Summative	Lesson Objective Assessed
5	Stakeholder Group Printed Materials *rubric*	Group	Summative	• Apply understanding of nuclear energy to create a stakeholder-specific response to a fictional nuclear accident. • Use persuasive language to present an argument to a target audience. • Apply the EDP to solve a complex problem. • Collaborate with peers to solve a problem.
5	Stakeholder Group Prototype Design *rubric*	Group	Summative	• Create a prototype related to a science or technology aspect of nuclear energy. • Apply the EDP to solve a complex problem. • Collaborate with peers to solve a problem.
5	STEM Research Notebook *prompt*	Individual	Formative	• Apply the EDP to solve a complex problem.

MODULE TIMELINE

Tables 3.6–3.10 (pp. 39–41) provide lesson timelines for each week of the module. The timelines are provided for general guidance only and are based on class times of approximately 45 minutes.

Table 3.6. STEM Road Map Module Schedule for Week One

Day 1	Day 2	Day 3	Day 4	Day 5
Lesson 1 *Putting Radioactivity to Work*	*Lesson 1* *Putting Radioactivity to Work*	*Lesson 1* *Putting Radioactivity to Work*	*Lesson 1* *Putting Radioactivity to Work*	*Lesson 1* *Putting Radioactivity to Work*
• Launch the module by introducing nuclear energy and radioactive decay. • Introduce module challenge, the Gammatown Crisis Challenge.	• Students work in groups to develop a timeline of discoveries related to and uses of radioactivity. • Students explore the concept of half-life as an exponential function.	• Students continue work on Radioactivity Timeline and exponential functions. • Introduce engineering design process and Nuclear Fission Model project.	• Students work in groups to create Radioactive Decay Chain Models.	• Student groups complete their Radioactive Decay Chain Models.

Table 3.7. STEM Road Map Module Schedule for Week Two

Day 6	Day 7	Day 8	Day 9	Day 10
Lesson 2 *Harnessing the Atom's Power*	*Lesson 2* *Harnessing the Atom's Power*	*Lesson 2* *Harnessing the Atom's Power*	*Lesson 2* *Harnessing the Atom's Power*	*Lesson 2* *Harnessing the Atom's Power*
• Introduce nuclear fission. • Students explore the energy output of an atomic bomb, expressing this output in a variety of units.	• Students work in teams to create schematic diagrams of pressurized water reactors (PWRs)	• Students complete PWR diagrams and begin work on Nuclear Fission Models. • Students calculate energy outputs of fission reactions.	• Students continue work on Nuclear Fission Models.	• Students complete Nuclear Fission Models. • Students begin work on thorium reactor jigsaw essay.

Table 3.8. STEM Road Map Module Schedule for Week Three

Day 11	Day 12	Day 13	Day 14	Day 15
Lesson 2 *Harnessing the Atom's Power*	*Lesson 3* *Nuclear Fusion: Harnessing the Power of the Stars*	*Lesson 3* *Nuclear Fusion: Harnessing the Power of the Stars*	*Lesson 3* *Nuclear Fusion: Harnessing the Power of the Stars*	*Lesson 3* *Nuclear Fusion: Harnessing the Power of the Stars*
• Students complete thorium reactor essays.	• Introduce nuclear fusion. • Students calculate the energy output of a fusion reaction as compared to a nuclear fission reaction.	• Students begin work on Nuclear Fusion Models. • Students begin creating a plan for a product powered by nuclear energy.	• Students continue work on Nuclear Fusion Models. • Students research current innovations in nuclear fusion technologies.	• Students complete Nuclear Fusion Models. • Students complete work on product plans.

Table 3.9. STEM Road Map Module Schedule for Week Four

Day 16	Day 17	Day 18	Day 19	Day 20
Lesson 4 *Anatomy of a Nuclear Accident*	*Lesson 4* *Anatomy of a Nuclear Accident*	*Lesson 4* *Anatomy of a Nuclear Accident*	*Lesson 4* *Anatomy of a Nuclear Accident*	*Lesson 5* *The Gammatown Crisis Challenge*
• Introduce the Three Mile Island (TMI) nuclear accident. • Students begin reading and analyzing an account of the accident. • Introduce measures of human exposure to radiation.	• Students continue reading and analyzing TMI accident account. • Students calculate radiation doses associated with various activities.	• Students continue reading and analyzing TMI accident account. • Student teams create topic-specific posters. • Students continue working with radiation emission and exposure calculations.	• Student teams complete and present topic-specific posters about the TMI accident. • Students continue working with radiation emission and exposure calculations.	• Review challenge requirements and assign team stakeholder groups. • Students work on Define step of EDP to create challenge solutions.

Table 3.10. STEM Road Map Module Schedule for Week Five

Day 21	Day 22	Day 23	Day 24	Day 25
Lesson 5 *The Gammatown Crisis Challenge* • Students work on Learn step of EDP to create challenge solutions.	*Lesson 5* *The Gammatown Crisis Challenge* • Students work on Plan step of EDP to create challenge solutions.	*Lesson 5* *The Gammatown Crisis Challenge* • Students work on Try step of EDP to create challenge solutions.	*Lesson 5* *The Gammatown Crisis Challenge* • Students work on Test and Decide steps of EDP to create challenge solutions.	*Lesson 5* *The Gammatown Crisis Challenge* • Students present stakeholder group–specific challenge solutions in a town hall meeting.

RESOURCES

The media specialist can help teachers locate resources for students to view and read about nuclear energy, the history of energy production, the development of nuclear weapons in the United States, and related physics and chemistry content. Special educators and reading specialists can help find supplemental sources for students needing extra support in reading and writing. Additional resources may be found online. Community resources for this module may include mechanical engineers, nuclear engineers, power plant representatives, specialists in occupational safety and health, and community emergency response coordinators.

REFERENCES

Keeley, P., and R. Harrington. 2010. *Uncovering student ideas in physical science, volume 1: 45 new force and motion assessment probes.* Arlington, VA: NSTA Press.

National Research Council (NRC). 1997. *Science teaching reconsidered: A handbook.* Washington, DC: National Academies Press.

Peters-Burton, E. E., P. Seshaiyer, S. R. Burton, J. Drake-Patrick, and C. C. Johnson. 2015. The STEM Road Map for grades 9–12. In *STEM Road Map: A framework for integrated STEM education*, ed. C. C. Johnson, E. E. Peters-Burton, and T. J. Moore, 124–162. New York: Routledge. *www.routledge.com/products/9781138804234.*

WIDA. 2012. 2012 amplification of the English language development standards: Kindergarten–grade 12. *https://wida.wisc.edu/teach/standards/eld.*

RADIOACTIVITY LESSON PLANS

Janet B. Walton, Bradley D. Rankin, Erin Peters-Burton, Anthony Pellegrino, Jennifer Drake-Patrick, and Carla C. Johnson

Lesson Plan 1: Putting Radioactivity to Work

This lesson provides a basic introduction to nuclear chemistry and exponential functions. Students learn about the discovery and history of radioactive elements and the scientific and societal implications of these discoveries. Students model natural radioactive decay as a background for understanding how radioactive elements are used in the controlled fission reactions used in nuclear reactors. Students also learn how to express half-lives of radioactive elements mathematically as exponential functions.

ESSENTIAL QUESTIONS

- What are the properties of radioactive elements that cause them to break down into different elements?

- What are the by-products of radioactive decay, and are they harmful to humans?

- How can radioactive decay be used for human purposes?

ESTABLISHED GOALS AND OBJECTIVES

At the conclusion of this lesson, students will be able to do the following:

- Discuss how radioactive decay was discovered and how scientific understanding of the process progressed to our present-day understanding

- Describe how the minuscule amount of energy released in a single alpha, beta, or gamma decay is significant because of the proportion of atoms in matter

- Discuss the major nuclear power plant accidents of the 20th and 21st centuries, and identify implications of these accidents

- Understand that radioactive decay is an exponential function

- Create real-world examples of exponential functions other than radioactive decay

- Model the decay chain of a radioactive element

- Identify several uses of radioactive elements and imagine ideas for future use

- Apply the engineering design process (EDP) to solve a problem

- Collaborate with peers to solve a problem

TIME REQUIRED

- 5 days (approximately 45 minutes each; see Table 3.6, p. 39)

MATERIALS

Required Materials for Lesson 1

- STEM Research Notebooks (1 per student; see p. 26 for STEM Research Notebook student handout)

- Computers with internet access for student research and viewing videos (for each team)

- Periodic tables (1 per student)

- Software for 3-D modeling (for teams who choose to create computer-simulated models of radioactive decay)

- Art supplies (for teams who choose to make physical models of radioactive decay)

- Handouts (attached at the end of this lesson)

Additional Materials for Sweetium Half-Life (per pair)

- 50 pieces of small candies with a letter on one side, such as M&M's or Skittles (*Note:* If you wish to allow students to eat candy afterward, purchase extra, as students will handle the candy in groups during the activity. Remind students not to eat food used in the lab.)

- Paper towel

- 12 oz. plastic cup

- Quart-size zipper seal bag

- 2 sheets of graph paper

- Safety goggles

Additional Materials for ELA Connection (optional)

- *Close Your Eyes, Hold Hands* by Chris Bohjalian (Doubleday, 2014; 1 per student)

SAFETY NOTES

1. All students must wear safety goggles during all phases of this inquiry activity.

2. Do not eat any food used during this investigation.

3. Pick up any candies that fall on the floor to avoid a slip-and-fall hazard.

4. Wash hands with soap and water after the activity is completed.

CONTENT STANDARDS AND KEY VOCABULARY

Table 4.1 lists the content standards from the *Next Generation Science Standards* (*NGSS*), *Common Core State Standards* (*CCSS*), and the Framework for 21st Century Learning that this lesson addresses, and Table 4.2 (p. 51) presents the key vocabulary. Vocabulary terms are provided for both teacher and student use. Teachers may choose to introduce some or all of the terms to students.

Table 4.1. Content Standards Addressed in STEM Road Map Module Lesson 1

NEXT GENERATION SCIENCE STANDARDS
PERFORMANCE EXPECTATIONS
• HS-PS1-1. Use the periodic table as a model to predict the relative properties of elements based on the patterns of electrons in the outermost energy level of atoms.
• HS-PS1-8. Develop models to illustrate the changes in the composition of the nucleus of the atom and the energy released during the processes of fission, fusion, and radioactive decay.
• HS-ETS-2. Design a solution to a complex real-world problem by breaking it down into smaller, more manageable problems that can be solved through engineering.
SCIENCE AND ENGINEERING PRACTICES
Using Mathematics and Computational Thinking
• Use mathematical representations of phenomena to support claims.
• Create a computational model or simulation of a phenomenon, designed device, process, or system.

Continued

Radioactivity, Grade 11

Table 4.1. (*continued*)

SCIENCE AND ENGINEERING PRACTICES (*continued*)

Developing and Using Models

- Use a model to predict the relationships between systems or between components of a system.

- Develop a model based on evidence to illustrate the relationships between systems or between components of a system.

Obtaining and Communicating Information

- Evaluate the validity and reliability of multiple claims that appear in scientific and technical texts or media reports, verifying the data when possible.

- Communicate technical information or ideas in multiple formats.

Constructing Explanations and Designing Solutions

- Design, evaluate, and/or refine a solution to a complex real-world problem, based on scientific knowledge, student-generated sources of evidence, prioritized criteria, and trade-off considerations.

- Construct and revise an explanation based on valid and reliable evidence obtained from a variety of sources (including students' own investigations, models, theories, simulations, peer review) and the assumption that theories and laws that describe the natural world operate today as they did in the past and will continue to do so in the future.

DISCIPLINARY CORE IDEAS

PS1.A: Structure and Properties of Matter

- Each atom has a charged substructure consisting of a nucleus, which is made of protons and neutrons, surrounded by electrons.

- The periodic table orders elements horizontally by the number of protons in the atom's nucleus and places those with similar chemical properties in columns. The repeating patterns of this table reflect patterns of outer electron states.

- A stable molecule has less energy than the same set of atoms separated; one must provide at least this energy in order to take the molecule apart.

PS1.B: Chemical Reactions

- The fact that atoms are conserved, together with knowledge of the chemical properties of the elements involved, can be used to describe and predict chemical reactions.

Continued

Table 4.1. (*continued*)

DISCIPLINARY CORE IDEAS (*continued*)

PS1.C: Nuclear Processes

- Nuclear processes, including fusion, fission, and radioactive decays of unstable nuclei, involve release or absorption of energy. The total number of neutrons plus protons does not change in any nuclear process.

- Spontaneous radioactive decays follow a characteristic exponential decay law. Nuclear lifetimes allow radiometric dating to be used to determine the ages of rocks and other materials.

PS3.A: Definitions of Energy

- Energy is a quantitative property of a system that depends on the motion and interactions of matter and radiation within that system. That there is a single quantity called energy is due to the fact that a system's total energy is conserved, even as, within the system, energy is continually transferred from one object to another and between its various possible forms.

PS3.B: Conservation of Energy and Energy Transfer

- Conservation of energy means that the total change of energy in any system is always equal to the total energy transferred into or out of the system.

- Energy cannot be created or destroyed, but it can be transported from one place to another and transferred between systems.

- Mathematical expressions, which quantify how the stored energy in a system depends on its configuration (e.g., relative positions of charged particles, compression of a spring) and how kinetic energy depends on mass and speed, allow the concept of conservation of energy to be used to predict and describe system behavior.

- The availability of energy limits what can occur in any system.

PS3.D: Energy in Chemical Processes

- Although energy cannot be destroyed, it can be converted to less useful forms—for example, to thermal energy in the surrounding environment.

- Nuclear fusion processes in the center of the sun release the energy that ultimately reaches Earth as radiation.

CROSSCUTTING CONCEPTS

Patterns

- Empirical evidence is needed to identify patterns.

Continued

Table 4.1. (*continued*)

CROSSCUTTING CONCEPTS (*continued*)

Energy and Matter

- Changes of energy and matter in a system can be described in terms of energy and matter flows into, out of, and within that system.

- In nuclear processes, atoms are not conserved, but the total number of protons plus neutrons is conserved.

- The total amount of energy and matter in closed systems is conserved.

Stability and Change

- Much of science deals with constructing explanations of how things change and how they remain stable.

Systems and System Models

- Models can be used to predict the behavior of a system, but these predictions have limited precision and reliability due to the assumptions and approximations inherent in models.

Structure and Function

- Investigating or designing new systems or structures requires a detailed examination of the properties of different materials, the structures of different components, and connections of components to reveal its function and/or solve a problem.

COMMON CORE STATE STANDARDS FOR MATHEMATICS

MATHEMATICAL PRACTICES

- MP1. Make sense of problems and persevere in solving them.

- MP3. Construct viable arguments and critique the reasoning of others.

- MP4. Model with mathematics.

- MP7. Look for and make use of structure.

MATHEMATICAL CONTENT

- HSA.APR.D.6. Rewrite simple rational expressions in different forms; write $a(x)/b(x)$ in the form $q(x) + r(x)/b(x)$, where $a(x)$, $b(x)$, $q(x)$, and $r(x)$ are polynomials with the degree of $r(x)$ less than the degree of $b(x)$ using inspection, long division, or, for the more complicated examples, a computer algebra system.

- HSF.IF.B.4. For a function that models a relationship between two quantities, interpret key features of graphs and tables in terms of the quantities, and sketch graphs showing key features given a verbal description of the relationship.

- HSF.IF.B.6. Calculate and interpret the average rate of change of a function (presented symbolically or as a table) over a specified interval. Estimate the rate of change from a graph.

Continued

Table 4.1. (*continued*)

MATHEMATICAL CONTENT (continued)

- HSF.LE.A.1. Distinguish between situations that can be modeled with linear functions and with exponential functions.

- HSF.LE.A.2. Construct linear and exponential functions, including arithmetic and geometric sequences, given a graph, a description of a relationship, or two input-output pairs (include reading these from a table).

- HSF.LE.A.3. Observe using graphs and tables that a quantity increasing exponentially eventually exceeds a quantity increasing linearly, quadratically, or (more generally) as a polynomial function.

- HSF.LE.A.4. For exponential models, express as a logarithm the solution to $ab^{ct} = d$ where a, c, and d are numbers and the base b is 2, 10, or e; evaluate the logarithm using technology.

- HSF.LE.B.5. Interpret the parameters in a linear or exponential function in terms of a context.

- HSF.BF.A.1. Write a function that describes a relationship between two quantities.

- HSF.BF.A.2. Write arithmetic and geometric sequences both recursively and with an explicit formula, use them to model situations, and translate between the two forms.

- HSF.BF.B.3. Identify the effect on the graph of replacing $f(x)$ by $f(x) + k$, $k\,f(x)$, $f(kx)$, and $f(x + k)$ for specific values of k (both positive and negative); find the value of k given the graphs. Experiment with cases and illustrate an explanation of the effects on the graph using technology. Include recognizing even and odd functions from their graphs and algebraic expressions for them.

- HSF.BF.B.4. Find inverse functions.

- HSF.BF.B.5. Understand the inverse relationship between exponents and logarithms and use this relationship to solve problems involving logarithms and exponents.

COMMON CORE STATE STANDARDS FOR ENGLISH LANGUAGE ARTS

READING STANDARDS

- RI.11-12.1. Cite strong and thorough textual evidence to support analysis of what the text says explicitly as well as inferences drawn from the text, including determining where the text leaves matters uncertain.

- RI.11-12.2. Determine two or more central ideas of a text and analyze their development over the course of the text, including how they interact and build on one another to provide a complex analysis; provide an objective summary of the text.

- RI.11-12.3. Analyze a complex set of ideas or sequence of events and explain how specific individuals, ideas, or events interact and develop over the course of the text.

- RI.11-12.4. Determine the meaning of words and phrases as they are used in a text, including figurative, connotative, and technical meanings; analyze how an author uses and refines the meaning of a key term or terms over the course of a text.

Continued

Table 4.1. (*continued***)**

> **READING STANDARDS (***continued***)**
>
> - RI.11-12.5. Analyze and evaluate the effectiveness of the structure an author uses in his or her exposition or argument, including whether the structure makes points clear, convincing, and engaging.
>
> - RI.11-12.7. Integrate and evaluate multiple sources of information presented in different media or formats (e.g., visually, quantitatively) as well as in words in order to address a question or solve a problem.
>
> - RI.11-12.8. Delineate and evaluate the reasoning in seminal U.S. texts, including the application of constitutional principles and use of legal reasoning and the premises, purposes, and arguments in works of public advocacy.
>
> **SPEAKING AND LISTENING STANDARDS**
>
> - SL.11-12.1. Initiate and participate effectively in a range of collaborative discussions (one-on-one, in groups, and teacher-led) with diverse partners on grades 11–12 topics, texts, and issues, building on others' ideas and expressing their own clearly and persuasively.
>
> - SL.11-12.2. Integrate multiple sources of information presented in diverse formats and media (e.g., visually, quantitatively, orally) in order to make informed decisions and solve problems, evaluating the credibility and accuracy of each source and noting any discrepancies among the data.
>
> - SL.11-12.3. Evaluate a speaker's point of view, reasoning, and use of evidence and rhetoric, assessing the stance, premises, links among ideas, word choice, points of emphasis, and tone used.
>
> - SL.11-12.4. Present information, findings, and supporting evidence, conveying a clear and distinct perspective, such that listeners can follow the line of reasoning, alternative or opposing perspectives are addressed, and the organization, development, substance, and style are appropriate to purpose, audience, and a range of formal and informal tasks.
>
> - SL.11-12.5. Make strategic use of digital media (e.g., textual, graphical, audio, visual, and interactive elements) in presentations to enhance understanding of findings, reasoning, and evidence and to add interest.
>
> - SL.11-12.6. Adapt speech to a variety of contexts and tasks, demonstrating a command of formal English when indicated or appropriate.
>
> **FRAMEWORK FOR 21ST CENTURY LEARNING**
>
> - Interdisciplinary Themes: Global Awareness; Financial, Economic, Business and Entrepreneurial Literacy; Civic Literacy; Environmental Literacy
>
> - Learning and Innovation Skills: Creativity and Innovation, Critical Thinking and Problem Solving, Communication and Collaboration
>
> - Information, Media, and Technology Skills: Information Literacy, Media Literacy, ICT Literacy
>
> - Life and Career Skills: Flexibility and Adaptability, Initiative and Self-Direction, Social and Cross-Cultural Skills, Productivity and Accountability, Leadership and Responsibility

Table 4.2. Key Vocabulary for Lesson 1

Key Vocabulary	Definition
alpha decay	a process in which an atom ejects a particle with two neutrons, two protons, and no electrons (identical to the nucleus of a helium atom); the particle is called an alpha particle
alpha particles	charged particles emitted from radioactive materials; these particles are potentially dangerous if inhaled or swallowed but can easily be blocked from penetrating materials
beta decay	a process in which a neutron in an atom changes to a proton or a proton changes into a neutron and emits electrons or positrons (beta particles)
beta particles	charged particles that are electrons or positrons are emitted from radioactive materials; these particles can penetrate skin but are easily blocked by a metal, plastic, or wood barrier
electron	a subatomic particle with a negative charge
electron volt	the work done on an electron in moving it through a potential difference of one volt
gamma decay	a process in which a nucleus that has undergone alpha or beta decay emits electromagnetic radiation in the form of gamma ray photons
gamma rays	high-energy radiation that can travel large distances at the speed of light; these rays are able to penetrate many materials, including skin
half-life	the amount of time necessary for an unstable element to diminish to one-half of its original quantity
ionizing radiation	radiation that has sufficient energy to break molecular bonds and displace electrons as it moves through material; may causes changes in living cells, including those of plants, animals, and people; includes alpha and beta particles, gamma rays, and neutrons
isotopes	different forms of the same element that contain the same number of protons but different numbers of neutrons in their nuclei
neutrons	nuclear particles that can travel at high speeds and can penetrate other materials, making them radioactive
positron	a subatomic particle with the same mass as an electron but with a positive charge
proton	a positively charged subatomic particle in an atom's nucleus
radioactive decay or radioactivity	a process in which an unstable nucleus loses energy in the form of alpha, beta, or gamma rays; since the element's nucleus changes, it becomes a different element or a different isotope of the same element
subatomic particle	particles that are smaller than an atom and found inside an atom
transmutation	a process in which an element changes into a different element through radioactive decay

TEACHER BACKGROUND INFORMATION

This lesson introduces the module and the final challenge by introducing the history and current uses of nuclear energy. The information in this section will help you engage your students in this lesson.

Nuclear Chemistry

Nuclear chemistry is a branch of chemistry that deals with changes in matter that originate in the nucleus of an atom. These nuclear changes have enormous implications for power generation, weapon development, and other technologies, since they have the capacity to release the vast amounts of energy stored inside atomic nuclei. This module focuses on the use of nuclear energy as an alternative to fossil fuels to meet society's energy demands. This has become an increasingly controversial topic since accidents such as the nuclear core meltdowns of the Fukushima Daiichi reactors in Japan in 2011 and the explosion at the Chernobyl Nuclear Power Plant in Ukraine in 1986.

Nuclear chemistry focuses on the subatomic particles in an atom's nucleus: protons and neutrons. Atomic nuclei can undergo reactions that change the atom's identity because the number of protons and neutrons in the atom change in the course of nuclear reactions. When an atom has an unstable nucleus (i.e., the binding energy is not strong enough to hold the nucleus together), it spontaneously emits particles and electromagnetic radiation, allowing it to move to a lower-energy state and become more stable; these elements are said to be radioactive since they undergo radioactive decay. You should be familiar with the processes of radioactive decay and the types of ionizing radiation (alpha particles, beta particles, gamma rays, X-rays, and neutrons). Figure 4.1 (see p. 67) illustrates the various types of ionizing radiation and their ability to penetrate substances. You should also know how to use exponential models to represent the amount of radioactive material remaining in a given amount of material at any point in its decay process.

The following resources may be useful:

- U.S. Nuclear Regulatory Commission's "Radiation Basics": *www.nrc.gov/about-nrc/radiation/health-effects/radiation-basics.html*

- "E = mc²: Energy From Radioactive Decay": *www.emc2-explained.info/Emc2/Decay.htm#.VQXLeMaocsk*

- "Some Nuclear Units": *http://hyperphysics.phy-astr.gsu.edu/hbase/nuclear/nucuni.html*

- "Radioactive Decay": *http://en.wikipedia.org/wiki/Radioactive_decay*

- "60 Minutes Fukushima Report" video: *www.youtube.com/watch?v=4n05BeqPp1w*

- "ABC News Nightline: Chernobyl Accident" video: *www.youtube.com/watch?v=w_uOSImPSi8*

- "The Future of Clean Nuclear Energy Is Coming" video: *www.youtube.com/watch?v=t7FvxN_gkt4*

- "Exponential Equations: Half-Life Applications" video: *www.youtube.com/watch?v=rQVzNynZ9vM*

History and Uses of Nuclear Energy

For this lesson, you should be familiar with the highlights of the history of nuclear energy from its initial discovery in the 1800s through current applications of nuclear energy to meet world energy needs. Information about the history and uses of nuclear power in the United States can be found at the following websites:

- U.S. Nuclear Regulatory Commission: *www.nrc.gov*

- U.S. Energy Information Administration and its associated education website, "Energy Kids": *www.eia.gov* and *www.eia.gov/kids/energy.cfm?page=nuclear_home-basics*

- International Atomic Energy Agency: *www.iaea.org*

- World Nuclear Association: *www.world-nuclear.org*

- Argonne National Laboratory, with nuclear energy resources for schools: *https://students.ne.anl.gov/schools/us/1*

Engineering Design Process

In this module, students are challenged to work in teams to complete a variety of tasks. They will use the engineering design process (EDP), the same process that professional engineers use in their work, to structure their group collaboration. A graphic representation of the EDP is provided at the end of this lesson (p. 66). It may be useful to post this in your classroom.

The EDP is a cyclical process consisting of a number of steps that are iterative in nature:

- *Define:* Identify the problem.

- *Learn:* Brainstorm ideas and conduct background research.

- *Plan:* Create a plan (e.g., by sketching designs, formulating processes).

- *Try:* Build (this may mean building an object or formulating a process or solution).

- *Test:* What works and what doesn't? What could you improve?

- *Decide:* Redesign to solve problems that came up in testing, and then retest.

The following website provides additional information about the EDP: *www.pbslearningmedia.org/resource/phy03.sci.engin.design.desprocess/what-is-the-engineering-design-process.*

Mathematical Modeling

The mathematics content in this lesson focuses on modeling. Modeling provides a way for students to connect mathematics content to real-world phenomena and solve problems using data, and it can support students' understanding of mathematical concepts. Mathematical modeling is the foundation for a variety of fields, such as engineering, medicine, economics, and finance, and allows practitioners to make predictions. Equations, decision-making processes, and graphs are all types of mathematical models. Models can be conceptual or quantitative, or both. Conceptual models visually display ideas about the relationships between variables (a flow chart, for example), while quantitative models (equations) take into account the dynamic nature of the variables involved. Student models in this module will combine both conceptual and quantitative elements. For more information about modeling, see the following resources:

- *www.corestandards.org/Math/Content/HSM*

- *www.gettingsmart.com/2016/08/mathematical-modeling-for-high-school-students*

- *www.americanscientist.org/blog/macroscope/5-reasons-to-teach-mathematical-modeling*

Besides making two- or three-dimensional models using supplies such as poster board and modeling clay, students can use a variety of technology tools for mathematical modeling, such as web-based applets, graphing software, drawing tools, calculators, dynamic geometry software, and spreadsheets. MATLAB and Simulink are commercially available mathematical modeling programs (see *www.mathworks.com/academia/highschool.html*).

Interactive Simulations

The PhET website at *https://phet.colorado.edu/en/simulations/category/chemistry* provides online interactive simulations of several module topics that you may wish to incorporate into this and other lessons, including alpha decay, beta decay, isotopes and atomic mass, and nuclear fission, as well as the "Radioactive Dating Game."

COMMON MISCONCEPTIONS

Students will have various types of prior knowledge about the concepts introduced in this lesson. Table 4.3 outlines some common misconceptions students may have concerning these concepts. Because of the breadth of students' experiences, it is not possible to anticipate every misconception that students may bring as they approach this lesson. Incorrect or inaccurate prior understanding of concepts can influence student learning in the future, however, so it is important to be alert to misconceptions such as those presented in the table.

Table 4.3. Common Misconceptions About the Concepts in Lesson 1

Topic	Student Misconception	Explanation
Linear and exponential functions	Linear and exponential functions both change by a constant number.	Linear functions change by a constant number, whereas exponential functions change by a constant rate.
	An exponential function always expresses growth (i.e., an exponential increase).	Exponential functions can express both growth and decay.
	An exponential growth or decay rate is always fast.	Exponential growth or decay is not necessarily fast. Consider the example of a bank account that earns 0.5% interest; the growth rate would be expressed as an exponential function, but this exponential growth would not be fast.
Nuclear energy	An atom of an element cannot change to become a different element.	In nuclear reactions, the number of protons in an atom can change; this results in the atom becoming a different element.
	Radiation causes cancer and therefore cannot be used in healthcare.	Although excessive radiation can be a cause of cancer, radiation can also destroy cancerous cells and is therefore used as a cancer treatment.
	Radiation is produced only by human-caused nuclear reactions.	We are surrounded by naturally occurring radiation. There are naturally occurring radioactive materials in Earth's crust and in the atmosphere.

PREPARATION FOR LESSON 1

Review the Teacher Background Information section (p. 52), assemble the materials for the lesson, make copies of the student handouts, and preview the videos recommended in the Learning Components section that follows. Be prepared to lead a discussion on the basic principles of radioactive elements as a means of introducing this lesson's topic. Have available a list of sources students can use to learn about the discovery, history, processes, and present-day uses of radioactive elements. Have your students set up their STEM Research Notebooks (see pp. 25–26 for discussion and student instruction handout).

LEARNING COMPONENTS
Introductory Activity/Engagement

Connection to the Challenge: Begin each day of this lesson by directing students' attention to the driving question for the module and challenge: How does the use of nuclear energy to meet our energy demands affect society? Hold a brief discussion of students' existing knowledge about nuclear energy, creating a class list of key ideas on chart paper.

Science Class: Show students a variety of images of uses of nuclear energy and nuclear events, including Figures 4.2–4.5 (pp. 68–69) and others you found while preparing for this lesson. You may wish to also show videos such as the news clips on Fukushima and Chernobyl listed in the Teacher Background Information section (see p. 52).

Hold a class discussion about nuclear energy, giving students the opportunity to discuss their opinions of nuclear energy as a power source and whether they believe it should be used. Use questions such as the following to facilitate discussion:

- What is nuclear energy?

- Is it safe?

- Is it clean?

- Is it a renewable energy source?

- Why was it first used?

- What can it do for humans in the future?

Introduce the module challenge by distributing copies of the Gammatown Crisis Challenge Overview and Narrative handouts (pp. 71–74) to each student and reading them as a class. Explain to students that they will be grouped in teams of four to work on the challenge and that each team will be assigned a distinct role for the challenge. Assign students to their teams.

Next, show students a video about nuclear energy, such as "The Future of Clean Nuclear Energy Is Coming" at *www.youtube.com/watch?v=t7FvxN_gkt4*. Then, have students work in their groups to create lists of knowledge they think they will need to address the challenge.

STEM Research Notebook Prompt

Have students record their team lists of knowledge needed to address the challenge in their STEM Research Notebooks.

After students have completed their lists, provide an overview of topics they will cover during the module and have them compare this with the lists they made in their notebooks. The topic overview should include the following:

- Atomic science (how is energy released from atomic nuclei?)
- Radioactivity (what is it and what are its effects on living creatures?)
- Radioactive decay (half-lives)
- History and uses of nuclear energy
- Nuclear fission and nuclear fusion
- Nuclear reactors and how they work
- Pros and cons of nuclear energy
- The future of nuclear energy
- Problem-solving like a professional using the EDP

Mathematics Class: Have students complete the activity below.

STEM Research Notebook Prompt

Have students create a STEM Research Notebook entry in which they provide a definition based on their understanding of the term *half-life* (e.g., they may have heard the term used in relation to carbon dating or nuclear waste storage).

Next, have students share their definitions with the class, and develop a class definition based on student ideas. From a science perspective, students may offer definitions such as "the time it takes for half of a radioactive substance to turn into a nonradioactive substance." Emphasize the concept of half-life as a model of exponential decay in which the rate of decay is directly proportional to the amount of the substance present.

ELA Connection: Have students read an article about the history of radioactivity in medicine, such as "The History of Radiation Use in Medicine," by Amy B. Reed, at *www.jvascsurg.org/article/S0741-5214(10)01727-1/fulltext#sec3*.

STEM Research Notebook Prompt

Focus students' attention on the liberal use of radioactive elements in the late 19th and early 20th centuries, and have students create a fictional advertisement for a radioactive "miracle elixir" in their STEM Research Notebooks.

Social Studies Connection: Hold a class discussion about the role Marie Curie played in advancing gender equality in the sciences. You may begin by having students analyze what Curie said about not wearing a bridal gown for her wedding: "I have no dress except the one I wear every day. If you are going to be kind enough to give me one, please let it be practical and dark so that I can put it on afterwards to go to the laboratory." Have students read an article about Marie Curie's role in promoting gender equality in the sciences, such as "Is Marie Sklodowska Curie Still a Good Role Model for Female Scientists at 150?" at *www.independent.co.uk/news/science/marie-sklodowska-curie-cancer-scientist-female-role-model-a8043316.html*, and have students share their opinions about whether Curie is a good current role model for women in science fields. Advance the discussion by asking students to compare challenges women may have faced in the sciences in Curie's time with the challenges that they may face now.

Activity/Exploration

Science Class: Hold a class discussion about nuclear science and radioactivity by beginning with the question, "What is radioactivity?" Develop a class definition for radioactivity. Have students identify radioactive elements on the periodic table (see *www.periodictable.com/Elements/Radioactive* for images and more information on elements). Introduce the idea that radioactive particles decay (or seek more stable energy states) by emitting alpha particles, beta particles, or gamma rays (see p. 52 for more information).

Radioactivity Timeline

Explain to students that the discovery of radioactivity and its eventual use for nuclear technologies were not due to just one person, but that the understanding of radioactivity and its uses has been evolving since the 1800s. Tell students that you are going to create a collaborative timeline of nuclear science. Divide students into groups of two or three, and assign each to research a person, team of persons, or topic related to the discovery, history, and early use of nuclear science, such as the following:

- Henri Becquerel

- Marie Curie

- Wilhelm Roentgen (or Röntgen)

- Ernest Rutherford and Frederick Soddy

- Otto Hahn, Lise Meitner, and Fritz Strassmann

- Enrico Fermi

- Francis Perrin (the physicist)

- The Manhattan Project

- The use of X-ray machines for consumer purposes and entertainment (e.g., in shoe stores, for hair removal, and as carnival displays)

STEM Research Notebook Prompt

Students should answer the following questions about their team's assigned topic in their STEM Research Notebooks, citing the sources of their information:

- When did the person or persons live (or when did this use or activity occur)?

- Where did the person or persons do their work (or where did this use or activity occur)?

- How did the person or persons (or the use or activity) contribute to nuclear science?

- How have other people been affected by the person or persons (or the use or activity)?

Then, each group should create a one-page summary of its findings with the name of the person or persons or the use or activity and the date or dates in large print at the top of the page. Underneath the heading, the group should provide a one- to two-sentence summary of the contributions of the person or persons or the use or activity. Have groups organize themselves chronologically to present their findings to the class, with the group with the earliest date presenting first. Have students make notes in their STEM Research Notebooks so that the above questions are answered for all presentations. After each group presents its findings, it should post its one-page summary on a classroom wall in chronological order to create a class timeline.

After the timeline is complete, ask students to note the most recent date. Ask if they think there have been advances in nuclear science since then, and hold a class discussion about late-20th- and 21st-century advances in nuclear science.

Radioactive Decay Chain Model

Introduce the concept of radioactive decay to students, reminding students to take notes in their STEM Research Notebooks. They should understand that unstable isotopes decay by emitting alpha particles, beta particles, and gamma radiation. Using students'

knowledge of the processes of radioactive decay, guide them through an example of a decay chain model, pointing out that most unstable atoms do not decay to their most stable forms in just one step, but rather undergo a series of decay events, sometimes called a decay chain or radioactive cascade. The thorium-232 decay chain is illustrated in Figure 4.6 (p. 70). (See student textbooks or online resources for other examples.) Thorium-232 (represented by the symbol ^{232}Th) is the only naturally occurring isotope of thorium and is unstable. It goes through a complex series of decays in which it emits alpha, beta, and gamma radiation.

Tell students that they are going to work in teams of two or three to create their own models of decay chains, using the EDP. These can be mathematical, computer-generated, or physical models using either technology or art supplies. The example of the decay chain in Figure 4.6 provides students with a conceptual basis. However, explain to students that the figure should not be considered a template for the type of model they will create; they should research other types of models and focus on creating a visual representation of a decay chain that demonstrates their knowledge and understanding of radioactive decay through the emission of alpha and beta particles. You may either allow student teams to research and choose unstable elements on which to base their decay chains or assign them elements such as neptunium-237, americium-241, or thorium-232. Students should incorporate a scale factor in the model, accurately comparing the element to a tangible object on Earth or in space (e.g., the Sun to the planets or a bug to the world). Provide students with copies of the Radioactive Decay Chain Model Rubric (pp. 76–77) to clarify expectations.

Introduce the EDP as a process by which engineers and other STEM professionals solve problems and accomplish complex tasks (see p. 53 for more information). Emphasize to students that just as engineers routinely work collaboratively, they will work in teams to create the models and to solve the final challenge. The EDP provides a framework for this group work. Give each student a copy of the EDP graphic (p. 66) and review the steps. As they plan and create their models, they should note their progress in their STEM Research Notebooks by creating an entry for each stage of the EDP.

Mathematics Class: Introduce the Sweetium Half-Life activity. Pair students to complete the activity.

Sweetium Half-Life

Provide each pair of students with 50 pieces of small candies marked with a letter on one side (to represent atoms), a paper towel, a plastic cup, and a zipper seal bag. Give each student a copy of the Sweetium Half-Life student handout (p. 75) and a piece of graph paper. Have student pairs follow the instructions on the handout to collect data to determine the half-life of sweetium. Ask the pairs to share their findings, and create a class chart of half-lives. Have students calculate the average half-life of sweetium for the class. Then, ask several students to share their graphs with the class. Discuss the shapes of the graphs (they should have the same basic shape). Hold a class discussion about why it is important to know the half-lives of substances. You may also wish to have students enter data from their tables into graphing calculators to determine the curve of best fit for their graphs and to generate an exponential regression equation in the form $y = a(b)^x$.

ELA Connection: A suggested literature connection is *Close Your Eyes, Hold Hands,* by Chris Bohjalian, a coming-of-age story about a 17-year-old girl who flees her home in Vermont after a meltdown at the local nuclear power plant for which her parents are blamed. Besides connecting to the topics of technical and social issues surrounding nuclear energy, nuclear power plants, and radiation, the novel also has literary connections to the work of Emily Dickinson and touches on social issues such as homelessness and child abuse. It also contains some references to prostitution and substance abuse. The publisher provides a reader's guide and sample discussion questions at *www.randomhouse.com/highschool/catalog/display.pperl?isbn=9780307743930&view=rg.*

Students may also search for primary resource data on the early applications of radioactive materials (for instance, in the Manhattan Project) and learn how to cite such data in APA and MLA formats. The Purdue Online Writing Lab (OWL) at *https://owl.purdue.edu/owl/purdue_owl.html* provides a comprehensive guide for using these and other citation methods. You may wish to have students use online citation tools such as Endnote or Zotero to aid them in the activity. Students should be encouraged to use appropriate citations as they research topics throughout the module.

Social Studies Connection: Show a brief video to introduce radiocarbon dating, such as "How Does Radiocarbon Dating Work?" at *www.youtube.com/watch?v=phZeE7Att_s.* After students watch the video, hold a class discussion on how radiocarbon dating affects society's understanding about the age of Earth and the impact that information has had on religious versus scientific views of Earth's origins. Have students consider how the advent of radiocarbon dating may have affected views of the past, religious origin stories, and people's views about the role of humans in the world's ecology.

In addition, you might have students use "The History of Nuclear Energy," by the U.S. Department of Energy, at *www.energy.gov/sites/prod/files/The%20History%20of%20Nuclear%20Energy_0.pdf,* to categorize events on the timeline as either advances or setbacks for nuclear energy. (*Note:* This publication includes history only until 1993 and therefore does not include more recent events such as the Fukushima Daiichi accident; have students research events after that date to categorize them.)

Explanation

Science Class: Students need to have a basic understanding of atomic structure, isotopes, radioactivity, radioactive decay, and the periodic table for this module. Introduce the concept of ionizing radiation and alpha particles, beta particles, gamma rays, and neutrons (see Table 4.2 on p. 51 for definitions). Show students Figure 4.1 (p. 67), a graphic of the types of ionizing radiation and their ability to penetrate substances. Introduce the concept of transmutation.

Mathematics Class: Students should understand that radioactive decay is an exponential function. As such, decay can be expressed as $A = A_0 \times (1/2)^{t/h}$, where A = final amount, A_0 = initial amount, t = time elapsed, and h = half-life. You may wish to provide your students with sample problems. For instance, if a radioactive isotope has a half-life of 500 years and you start with a sample of 50 kilograms (kg), how much will be left in 100 years?

To accommodate the varying needs of algebra II, mathematics analysis, and precalculus students, you may teach half-life at varying levels of complexity. For example, algebra II teachers can have students solve problems in which the half-life of a radioactive element is provided and they must compare the amount of the element to a given time in its decay model on a coordinate plane. Mathematics analysis teachers may wish to have students create an exponential model given the decay rate (k) of an element and use logarithms to predict the time it will take for half of the element to decay into a more stable element. Precalculus teachers can use the same scenario but give students time intervals and remaining amounts of an element and have them calculate the decay rate. In all classes, students should note the differences in the rates of change of each element modeled and make conjectures as to why such differences exist.

ELA Connection: Students will use persuasive arguments and informational speech and writing in the final challenge. You may wish to provide examples of various texts (e.g., advertisements, public service announcements, newspaper articles, speeches by public officials) that illustrate these uses of language.

Social Studies Connection: Not applicable.

Elaboration/Application of Knowledge

Science Class: Have student teams present their models of radioactive decay to the class.

Mathematics Class: Tell students that there are other processes besides radioactive decay that reflect exponential decay or growth. Examples include annual interest rates in bank accounts, bacterial growth, and elimination rounds in sports contests.

STEM Research Notebook Prompt

Have students respond to the following prompt in their STEM Research Notebooks: *What processes other than radioactive decay reflect exponential decay or growth? You may think of examples in science or in other areas such as finance or sports. Describe two of these processes and explain how their growth or decay is exponential. Create a problem based on one of these processes for the class to solve and graph. Be sure to provide a solution to the problem, including a graph.*

ELA Connection: Have students prepare for a debate on nuclear energy to be held at the end of the module. Assign students a stance either for or against nuclear energy. They should conduct research on each side of the issue both to compile supporting points for their own arguments and to be prepared to respond to the opposing side's arguments.

Social Studies Connection: Have students compare the versions of the origins of Earth documented in various sources including their science textbooks and internet resources. Students should compare and analyze the various explanations, identifying discrepancies, omissions, and biases. Ask students to describe the level of detail they find in their textbook versus internet resources.

Evaluation/Assessment

Students may be assessed on the following performance tasks and other measures listed.

Performance Tasks

- Radioactive Decay Chain Model Rubric (pp. 76–77)
- Radioactivity Timeline

Other Measures

- STEM Research Notebook entries
- Evidence of using the EDP for the Radioactive Decay Chain Model

INTERNET RESOURCES

U.S. Nuclear Regulatory Commission's "Radiation Basics"
- *www.nrc.gov/about-nrc/radiation/health-effects/radiation-basics.html*

"E = mc²: Energy From Radioactive Decay"
- *www.emc2-explained.info/Emc2/Decay.htm#.VQXLeMaocsk*

"Some Nuclear Units"
- *http://hyperphysics.phy-astr.gsu.edu/hbase/nuclear/nucuni.html*

"Radioactive Decay"
- *http://en.wikipedia.org/wiki/Radioactive_decay*

"60 Minutes Fukushima Report" video
- *www.youtube.com/watch?v=4n05BeqPp1w*

"ABC News Nightline: Chernobyl Accident" video
- *www.youtube.com/watch?v=w_uOSImPSi8*

"The Future of Clean Nuclear Energy Is Coming" video
- *www.youtube.com/watch?v=t7FvxN_gkt4*

"Exponential Equations: Half-Life Applications" video
- *www.youtube.com/watch?v=rQVzNynZ9vM*

U.S. Nuclear Regulatory Commission
- *www.nrc.gov*

U.S. Energy Information Administration and its associated education website, "Energy Kids"
- *www.eia.gov*

- *www.eia.gov/kids/energy.cfm?page=nuclear_home-basics*

International Atomic Energy Agency
- *www.iaea.org*

World Nuclear Association
- *www.world-nuclear.org*

Argonne National Laboratory, with nuclear energy resources for schools
- *https://students.ne.anl.gov/schools/us/1*

Information about the EDP
- *www.pbslearningmedia.org/resource/phy03.sci.engin.design.desprocess/what-is-the-engineering-design-process*

Mathematical modeling
- *www.corestandards.org/Math/Content/HSM*

- *www.gettingsmart.com/2016/08/mathematical-modeling-for-high-school-students*

- *www.americanscientist.org/blog/macroscope/5-reasons-to-teach-mathematical-modeling*

Mathematical modeling programs
- *www.mathworks.com/academia/highschool.html*

Interactive simulations
- *https://phet.colorado.edu/en/simulations/category/chemistry*

"The History of Radiation Use in Medicine," by Amy B. Reed
- *www.jvascsurg.org/article/S0741-5214(10)01727-1/fulltext#sec3*

"Is Marie Sklodowska Curie Still a Good Role Model for Female Scientists at 150?"
- *www.independent.co.uk/news/science/marie-sklodowska-curie-cancer-scientist-female-role-model-a8043316.html*

Radioactive elements on the periodic table
- *www.periodictable.com/Elements/Radioactive*

Reader's guide and discussion questions for *Close Your Eyes, Hold Hands,* by Chris Bohjalian
- *www.randomhouse.com/highschool/catalog/display.pperl?isbn=9780307743930&view=rg*

Purdue Online Writing Lab (OWL)
- *https://owl.purdue.edu/owl/purdue_owl.html*

"How Does Radiocarbon Dating Work?" video
- *www.youtube.com/watch?v=phZeE7Att_s*

"The History of Nuclear Energy," by the U.S. Department of Energy
- *www.energy.gov/sites/prod/files/The%20History%20of%20Nuclear%20Energy_0.pdf*

ENGINEERING DESIGN PROCESS

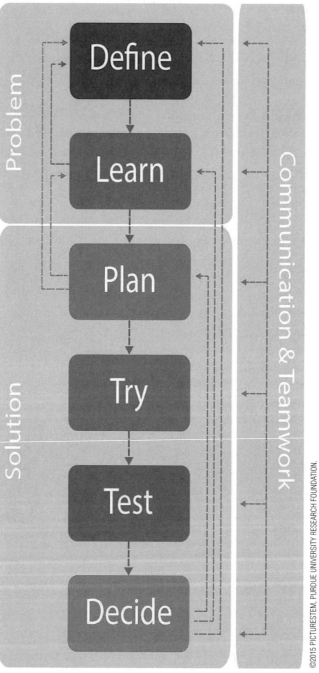

NATIONAL SCIENCE TEACHERS ASSOCIATION

FIGURE 4.1. TYPES OF RADIATION AND PENETRATION

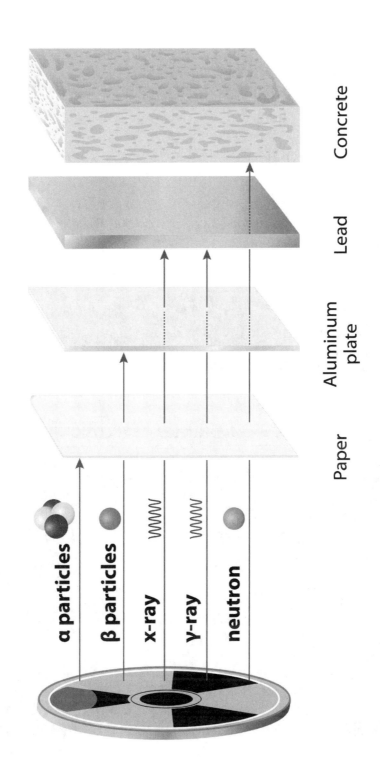

Note. A full-color version of this image is available on the book's Extras page at *www.nsta.org/roadmap-radioactivity.*

FIGURE 4.2. AN OPERATING NUCLEAR POWER PLANT

FIGURE 4.3. NUCLEAR EXPLOSION

Note: Full-color versions of these images are available on the book's Extras page at *www.nsta.org/roadmap-radioactivity.*

FIGURE 4.4. ANTINUCLEAR PROTEST IN JAPAN

FIGURE 4.5. USES OF NUCLEAR ENERGY

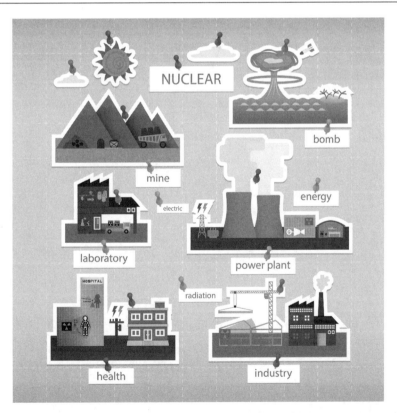

Note: Full-color versions of these images are available on the book's Extras page at *www.nsta.org/roadmap-radioactivity.*

FIGURE 4.6. THE THORIUM-232 DECAY CHAIN

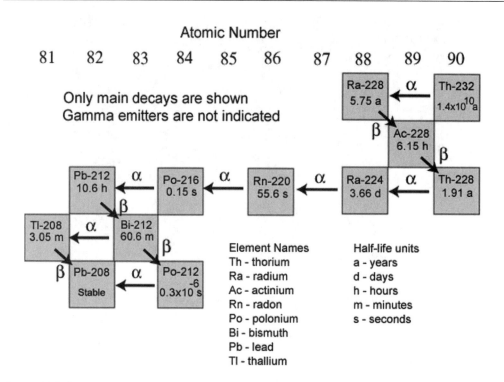

Source: U.S. Geological Survey at *http://pubs.usgs.gov/of/2004/1050/thorium.htm* (public domain).

GAMMATOWN CRISIS CHALLENGE OVERVIEW

Gammatown is in trouble and needs YOUR help! This charming town in the northeastern United States is in crisis. The local nuclear power plant's reactor #1 experienced a partial meltdown last week, and you and your team are charged with responding to that crisis. Your team will take on a role as a stakeholder involved in the Gammatown incident of 2019.

Your group will act as one of the following stakeholder groups and will prepare materials for your target audience:

✓ Public health officials educating the public about health risks

✓ A public relations team from the power plant providing educational material to news media

✓ Power plant managers trying to persuade the Nuclear Regulatory Commission (NRC) to continue to allow the reactor to operate

✓ NRC officials investigating the causes of the accident and preparing a report for public officials, including the governor and the president

✓ A team from the Environmental Protection Agency assessing risk to local natural resources and wildlife

✓ A group of local antinuclear activists who have been lobbying for the Gammatown nuclear plant to close since it opened

✓ A group of local business owners and economic development officials seeking to maintain the growing tourism industry, who are in the midst of closing a deal with an automobile seat belt manufacturer to build a plant outside of town that will employ 500 people

Your group will be responsible for the following:

✓ A presentation in the media of your choice (e.g., video, public service announcement, PowerPoint, oral presentation with visual aids) that uses historical, mathematical, and scientific facts in a persuasive manner; you will present this in a town hall meeting, which will be a public forum with an opportunity for attendees to ask questions

STUDENT HANDOUT, PAGE 2

GAMMATOWN CRISIS CHALLENGE OVERVIEW

- ✓ Printed material appropriate for your target audience (e.g., brochure, white paper, billboard, flyers)

- ✓ A model related to the science or technology aspect most relevant to your position (this may be a physical, mathematical, or computer-designed model)

You and your group will be assessed on the following:

- ✓ The appropriateness of your materials to your target audience

- ✓ Your understanding of the scientific and historical concepts associated with nuclear power

- ✓ The accuracy and innovativeness of your model

- ✓ Your ability to answer questions related to your materials and your presentation

- ✓ Evidence that your team used the engineering design process (EDP) in creating your materials

- ✓ Your collaboration in working with your group

NATIONAL SCIENCE TEACHERS ASSOCIATION

GAMMATOWN CRISIS CHALLENGE NARRATIVE

Gammatown is a picturesque small town in the northeastern United States situated along the banks of the Scenic River. The town prides itself on its refurbished Main Street, populated by small, locally owned shops and restaurants, and on the large computer chip manufacturing plant that recently located to a business park on the outskirts of town. The economic downturn of 2007–08 was difficult for the town, but the economy experienced a resurgence in recent years, and the town is attracting increasing numbers of tourists and visitors from the nearby state capital, Green City.

Gammatown is especially important to the state because of the Gammatown Power Company nuclear reactor, five miles upstream of downtown on the Scenic River. The plant is slated to be closed when its operating license expires in 2020; however, officials had hoped to extend the license because of the plant's impeccable safety record and its reliable power generation since it opened in 1970. This pressurized water reactor (PWR) plant has two reactors on site with a combined generating capacity of 2,000 megawatts (MW). When the plant operates at 90% capacity, it supplies electricity for about 1.5 million households in the region, including Green City. The plant employs 800 people in this small northeastern town, and 25,000 people live within a five-mile radius of the plant.

One bleak spot in Gammatown's recent history is increasing reports of prescription drug abuse and addiction among the town's population. Health officials have suggested that this may be related to the high-stress work environment of nuclear power plant employees. While there is no clear evidence that this abuse is related to work in the power plant, and no unlawful activities have been detected among plant workers, there are rumors that several control room operators may abuse alcohol and prescription drugs.

On June 2, 2019, catastrophe struck when the Gammatown Power Company's reactor #1 suffered a partial meltdown of its uranium fuel core because of a malfunction in the cooling circuit. The malfunction was not detected until enough primary coolant had drained that the reactor core began to melt. Before the problem was corrected and the core was adequately cooled, radioactive particles had been released into the air and water. In particular, trace amounts of iodine-131 and caesium-137 were released into the atmosphere around the plant. The International Atomic Energy Agency assigned a preliminary rating to the event of 4 on the International Nuclear Event Scale. All residents within a two-mile radius of the plant were evacuated, and many more voluntarily chose to leave their homes. The accident occurred one week ago, and residents have been told that they can return to their homes in another week, although many are unwilling to do so.

STUDENT HANDOUT, PAGE 2

GAMMATOWN CRISIS CHALLENGE NARRATIVE

The Gammatown reactor has become a staple of the nightly news, warranting its own theme songs on various networks. These news reports are accompanied by menacing images, including a lone concrete cooling tower under gray skies, residents fleeing their homes with cars packed full of personal belongings, and the frightened eyes of children being herded onto school buses to evacuate on the day of the accident. Misinformation is rampant, and there have been unverified reports of dead fish in the Scenic River, squirrels running frantically around Gammatown Park colliding head-on with trees, and even one image of a sidewalk puddle with an unearthly green glow emanating from it.

Name: _____

STUDENT HANDOUT

SWEETIUM HALF-LIFE

In this activity, each piece of candy represents an atom of the radioactive element sweetium. You will determine the half-life for your sweetium using the procedure below.

1. Place 50 atoms of sweetium into a zipper seal bag.
2. Seal the bag and gently shake it for about 10 seconds.
3. Gently pour the sweetium atoms out onto the paper towel.
4. Count the number of pieces with the letter side facing up. This is the number of atoms that have decayed. Record this number on the data chart provided.
5. Place the sweetium atoms that have decayed (those with the letter side up) into the plastic cup.
6. Record the number of undecayed atoms (those with the letter side down) on the data chart.
7. Return the undecayed atoms to the bag and reseal the bag.
8. Repeat steps 1–7 until all but 3 or 4 of your sweetium atoms have decayed.
9. Graph the number of undecayed sweetium atoms versus time (assume that each trial = 1 year).
10. Create a graph of the number of decayed sweetium atoms over time.

Trial	Number of Atoms Decayed	Number of Undecayed Atoms Remaining
0		
1		
2		
3		
4		
5		
6		
7		
8		
9		
10		
11		
12		
13		
14		

What was the half-life of your sweetium? _____

Radioactive Decay Chain Model Rubric

Team Name: _____

Type of Model	Needs Improvement (1 point)	Approaches Expectations (2 points)	Meets Expectations (3 points)	Exceeds Expectations (4 points)	Score
PHYSICAL 3-D MODEL	• The model is missing major components and does not represent radioactive decay correctly.	• The model includes the correct initial element but is lacking the decay element or radiation/energy.	• The model is created from 3-D objects and includes the nucleons and electrons of the initial element, but with a few minor errors, as well as the decay element or radiation/energy.	• The model is created from 3-D objects and includes the nucleons and electrons of the initial element, as well as the decay element or radiation/energy.	
COMPUTER-GENERATED MODEL	• The model is a single computer-generated picture of the decay process. • There are several mistakes.	• The model demonstrates the decay process through computer-generated images or frames that are essentially pictures. • There are mistakes in the model.	• The model aptly demonstrates the decay process through some form of stop-motion animation (not fluid) that shows an element breaking down over time. • The initial model is accurate with few flaws.	• The model aptly demonstrates the decay process through some form of fluid animation that shows an element breaking down over time. • The initial model is accurate and without flaws.	

Continued

Radioactive Decay Chain Model Rubric (*continued*)

Type of Model	Needs Improvement (1 point)	Approaches Expectations (2 points)	Meets Expectations (3 points)	Exceeds Expectations (4 points)	Score
MATHEMATICAL MODEL	• The model shows only the change in mass. • There are several errors in the mathematics.	• The mathematical model describes the change in mass of the decay process without including the energy yield calculations.	• The mathematical model describes the change in mass and the energy yield in the decay process with some minor mistakes.	• A simple drawing of a decay process or copy of a model from the internet is accompanied by mathematical models or calculations actly describing the change in the mass of a decaying element as well as the energy yield. • The model also calculates the total energy yield of a 1 gram portion of the element.	
2-D MODEL	• The model is inaccurate, and the scale factor to a real-life object is not included.	• The model is created well, but the scale factor to a real-life object is not included.	• The model is created well but has some errors in the mathematics describing the scale factor to a real-life object.	• Model is hand-drawn or created with computer software. It is well organized, describes the process to the layman, and incorporates a scale factor, accurately comparing the element to a real-life object.	

TOTAL SCORE: _____

COMMENTS:

Lesson Plan 2: Harnessing the Atom's Power

In this lesson, students explore the process of nuclear fission from historical, scientific, and environmental perspectives. Students focus on how nuclear fission is used in pressurized water reactors (PWRs), the type of nuclear reactors most commonly used in the United States today. Students learn about the history of human use of nuclear fission and its societal and ecological ramifications and create a fission model.

ESSENTIAL QUESTIONS

- What is nuclear fission?

- How is fission used to produce energy?

- What are the societal and environmental implications of using nuclear fission to meet society's energy demands?

ESTABLISHED GOALS AND OBJECTIVES

At the conclusion of this lesson, students will be able to do the following:

- Apply their understanding of the history of nuclear science to understand the current scientific and environmental context of using nuclear power to meet society's energy needs

- Understand that fission is a nuclear process in which a large atomic nucleus splits into smaller nuclei, releasing the energy stored in the nuclear bonds

- Understand that some fission reactions occur spontaneously, while others, such as those used in nuclear reactors, require an energy input

- Understand that a nuclear chain reaction occurs when neutrons released during fission cause fission in one or more other nuclei

- Apply their understanding of fission and chain reactions to an understanding of the operation of a nuclear reactor

- Understand that there are ecological implications of using fission reactions to produce electricity

- Identify thorium as an alternative to uranium for nuclear fission reactions and discuss the advantages and disadvantages of using thorium in nuclear reactors

- Understand that radioactive decay is an exponential function

- Create real-world examples of exponential functions other than radioactive decay

- Apply the EDP to solve a complex problem

- Collaborate with peers to solve a problem

TIME REQUIRED

- 6 days (approximately 45 minutes each day; see Tables 3.7 and 3.8, pp. 39–40)

MATERIALS

Required Materials for Lesson 2

- STEM Research Notebooks
- Computers with internet access for student research and viewing videos (for each team)
- Periodic tables (1 per student)
- Large sheet of paper or poster board for nuclear reactor diagrams (1 per team)
- Set of markers (1 per team)
- Ruler (1 per team)
- Software for 3-D modeling (for teams who choose to create computer-simulated models of nuclear fission)
- Art supplies (for teams who choose to make physical models of nuclear fission)
- Indirectly vented chemical splash safety goggles
- Handouts (attached at the end of this lesson)

SAFETY NOTES

1. All students must wear safety goggles during all phases of this inquiry activity.
2. Use caution when working with sharps to avoid cutting or puncturing skin.
3. Wash hands with soap and water after the activity is completed.

CONTENT STANDARDS AND KEY VOCABULARY

Table 4.4 (p. 80) lists the content standards from the *NGSS, CCSS,* and the Framework for 21st Century Learning that this lesson addresses, and Table 4.5 (p. 86) presents the key vocabulary. Vocabulary terms are provided for both teacher and student use. Teachers may choose to introduce some or all of the terms to students.

Table 4.4. Content Standards Addressed in STEM Road Map Module Lesson 2

NEXT GENERATION SCIENCE STANDARDS

PERFORMANCE EXPECTATIONS

- HS-PS1-1. Use the periodic table as a model to predict the relative properties of elements based on the patterns of electrons in the outermost energy level of atoms.

- HS-PS1-7. Use mathematical representations to support the claim that atoms, and therefore mass, are conserved during a chemical reaction.

- HS-PS1-8. Develop models to illustrate the changes in the composition of the nucleus of the atom and the energy released during the processes of fission, fusion, and radioactive decay.

- HS-ETS-2. Design a solution to a complex real-world problem by breaking it down into smaller, more manageable problems that can be solved through engineering.

SCIENCE AND ENGINEERING PRACTICES

Using Mathematics and Computational Thinking

- Use mathematical representations of phenomena to support claims.

- Create a computational model or simulation of a phenomenon, designed device, process, or system.

Developing and Using Models

- Use a model to predict the relationships between systems or between components of a system.

- Develop a model based on evidence to illustrate the relationships between systems or between components of a system.

Obtaining and Communicating Information

- Evaluate the validity and reliability of multiple claims that appear in scientific and technical texts or media reports, verifying the data when possible.

- Communicate technical information or ideas in multiple formats.

Constructing Explanations and Designing Solutions

- Design, evaluate, and/or refine a solution to a complex real-world problem, based on scientific knowledge, student-generated sources of evidence, prioritized criteria, and trade-off considerations.

- Construct and revise an explanation based on valid and reliable evidence obtained from a variety of sources (including students' own investigations, models, theories, simulations, peer review) and the assumption that theories and laws that describe the natural world operate today as they did in the past and will continue to do so in the future.

Continued

Table 4.4. (*continued*)

DISCIPLINARY CORE IDEAS

PS1.A: Structure and Properties of Matter

- Each atom has a charged substructure consisting of a nucleus, which is made of protons and neutrons, surrounded by electrons.

- The periodic table orders elements horizontally by the number of protons in the atom's nucleus and places those with similar chemical properties in columns. The repeating patterns of this table reflect patterns of outer electron states.

- A stable molecule has less energy than the same set of atoms separated; one must provide at least this energy in order to take the molecule apart.

PS1.B: Chemical Reactions

- The fact that atoms are conserved, together with knowledge of the chemical properties of the elements involved, can be used to describe and predict chemical reactions.

PS1.C: Nuclear Processes

- Nuclear processes, including fusion, fission, and radioactive decays of unstable nuclei, involve release or absorption of energy. The total number of neutrons plus protons does not change in any nuclear process.

- Spontaneous radioactive decays follow a characteristic exponential decay law. Nuclear lifetimes allow radiometric dating to be used to determine the ages of rocks and other materials.

PS2.B: Types of Interactions

- Attraction and repulsion between electric charges at the atomic scale explain the structure, properties, and transformations of matter, as well as the contact forces between material objects.

PS3.A: Definitions of Energy

- Energy is a quantitative property of a system that depends on the motion and interactions of matter and radiation within that system. That there is a single quantity called energy is due to the fact that a system's total energy is conserved, even as, within the system, energy is continually transferred from one object to another and between its various possible forms.

Continued

Table 4.4. (*continued*)

PS3.B: Conservation of Energy and Energy Transfer

- Conservation of energy means that the total change of energy in any system is always equal to the total energy transferred into or out of the system.

- Energy cannot be created or destroyed, but it can be transported from one place to another and transferred between systems.

- Mathematical expressions, which quantify how the stored energy in a system depends on its configuration (e.g., relative positions of charged particles, compression of a spring) and how kinetic energy depends on mass and speed, allow the concept of conservation of energy to be used to predict and describe system behavior.

- The availability of energy limits what can occur in any system.

PS3.D: Energy in Chemical Processes

- Although energy cannot be destroyed, it can be converted to less useful forms—for example, to thermal energy in the surrounding environment.

- Nuclear fusion processes in the center of the sun release the energy that ultimately reaches Earth as radiation.

ETS1.A: Defining and Delimiting an Engineering Problem

- Criteria and constraints also include satisfying any requirements set by society, such as taking issues of risk mitigation into account, and they should be quantified to the extent possible and stated in such a way that one can tell if a given design meets them.

ETS1.C: Optimizing the Design Solution

- Criteria may need to be broken down into simpler ones that can be approached systematically, and decisions about the priority of certain criteria over others (trade-offs) may be needed.

CROSSCUTTING CONCEPTS

Patterns

- Different patterns may be observed at each of the scales at which a system is studied and can provide evidence for causality in explanations of phenomena.

Energy and Matter

- Changes of energy and matter in a system can be described in terms of energy and matter flows into, out of, and within that system.

- In nuclear processes, atoms are not conserved, but the total number of protons plus neutrons is conserved.

- The total amount of energy and matter in closed systems is conserved.

Stability and Change

- Much of science deals with constructing explanations of how things change and how they remain stable.

Continued

Table 4.4. (*continued*)

CROSSCUTTING CONCEPTS (*continued*)

Cause and Effect

- Cause and effect relationships can be suggested and predicted for complex natural and human designed systems by examining what is known about smaller scale mechanisms within the system.

Scale, Proportion, and Quantity

- The significance of a phenomenon is dependent on the scale, proportion, and quantity at which it occurs.

Systems and System Models

- Models can be used to predict the behavior of a system, but these predictions have limited precision and reliability due to the assumptions and approximations inherent in models.

Structure and Function

- Investigating or designing new systems or structures requires a detailed examination of the properties of different materials, the structures of different components, and connections of components to reveal its function and/or solve a problem.

COMMON CORE STATE STANDARDS FOR MATHEMATICS

MATHEMATICAL PRACTICES

- MP1. Make sense of problems and persevere in solving them.

- MP3. Construct viable arguments and critique the reasoning of others.

- MP4. Model with mathematics.

- MP7. Look for and make use of structure.

MATHEMATICAL CONTENT

- HSA.APR.D.6. Rewrite simple rational expressions in different forms; write $a(x)/b(x)$ in the form $q(x) + r(x)/b(x)$, where $a(x)$, $b(x)$, $q(x)$, and $r(x)$ are polynomials with the degree of $r(x)$ less than the degree of $b(x)$ using inspection, long division, or, for the more complicated examples, a computer algebra system.

- HSF.IF.B.4. For a function that models a relationship between two quantities, interpret key features of graphs and tables in terms of the quantities, and sketch graphs showing key features given a verbal description of the relationship.

- HSF.IF.B.6. Calculate and interpret the average rate of change of a function (presented symbolically or as a table) over a specified interval. Estimate the rate of change from a graph.

- HSF.LE.A.1. Distinguish between situations that can be modeled with linear functions and with exponential functions.

Continued

Table 4.4. (*continued*)

MATHEMATICAL CONTENT (*continued*)

- HSF.LE.A.2. Construct linear and exponential functions, including arithmetic and geometric sequences, given a graph, a description of a relationship, or two input-output pairs (include reading these from a table).

- HSF.LE.A.3. Observe using graphs and tables that a quantity increasing exponentially eventually exceeds a quantity increasing linearly, quadratically, or (more generally) as a polynomial function.

- HSF.LE.A.4. For exponential models, express as a logarithm the solution to $ab^{ct} = d$ where a, c, and d are numbers and the base b is 2, 10, or e; evaluate the logarithm using technology.

- HSF.LE.B.5. Interpret the parameters in a linear or exponential function in terms of a context.

- HSF.BF.A.1. Write a function that describes a relationship between two quantities.

- HSF.BF.A.2. Write arithmetic and geometric sequences both recursively and with an explicit formula, use them to model situations, and translate between the two forms.

- HSF.BF.B.3. Identify the effect on the graph of replacing $f(x)$ by $f(x) + k$, $k\,f(x)$, $f(kx)$, and $f(x + k)$ for specific values of k (both positive and negative); find the value of k given the graphs. Experiment with cases and illustrate an explanation of the effects on the graph using technology. Include recognizing even and odd functions from their graphs and algebraic expressions for them.

- HSF.BF.B.4. Find inverse functions.

- HSF.BF.B.5. Understand the inverse relationship between exponents and logarithms and use this relationship to solve problems involving logarithms and exponents.

COMMON CORE STATE STANDARDS FOR ENGLISH LANGUAGE ARTS

READING STANDARDS

- RI.11-12.1. Cite strong and thorough textual evidence to support analysis of what the text says explicitly as well as inferences drawn from the text, including determining where the text leaves matters uncertain.

- RI.11-12.2. Determine two or more central ideas of a text and analyze their development over the course of the text, including how they interact and build on one another to provide a complex analysis; provide an objective summary of the text.

- RI.11-12.3. Analyze a complex set of ideas or sequence of events and explain how specific individuals, ideas, or events interact and develop over the course of the text.

- RI.11-12.4. Determine the meaning of words and phrases as they are used in a text, including figurative, connotative, and technical meanings; analyze how an author uses and refines the meaning of a key term or terms over the course of a text.

Continued

Table 4.4. (*continued*)

> **READING STANDARDS** (*continued*)
>
> - RI.11-12.5. Analyze and evaluate the effectiveness of the structure an author uses in his or her exposition or argument, including whether the structure makes points clear, convincing, and engaging.
>
> - RI.11-12.7. Integrate and evaluate multiple sources of information presented in different media or formats (e.g., visually, quantitatively) as well as in words in order to address a question or solve a problem.
>
> - RI.11-12.8. Delineate and evaluate the reasoning in seminal U.S. texts, including the application of constitutional principles and use of legal reasoning and the premises, purposes, and arguments in works of public advocacy.
>
> **WRITING STANDARDS**
>
> - W.11-12.1. Write arguments to support claims in an analysis of substantive topics or texts, using valid reasoning and relevant and sufficient evidence.
>
> - W.11-12.2. Write informative/explanatory texts to examine and convey complex ideas, concepts, and information clearly and accurately through the effective selection, organization, and analysis of content.
>
> **SPEAKING AND LISTENING STANDARDS**
>
> - SL.11-12.1. Initiate and participate effectively in a range of collaborative discussions (one-on-one, in groups, and teacher-led) with diverse partners on grades 11–12 topics, texts, and issues, building on others' ideas and expressing their own clearly and persuasively.
>
> - SL.11-12.2. Integrate multiple sources of information presented in diverse formats and media (e.g., visually, quantitatively, orally) in order to make informed decisions and solve problems, evaluating the credibility and accuracy of each source and noting any discrepancies among the data.
>
> - SL.11-12.3. Evaluate a speaker's point of view, reasoning, and use of evidence and rhetoric, assessing the stance, premises, links among ideas, word choice, points of emphasis, and tone used.
>
> - SL.11-12.4. Present information, findings, and supporting evidence, conveying a clear and distinct perspective, such that listeners can follow the line of reasoning, alternative or opposing perspectives are addressed, and the organization, development, substance, and style are appropriate to purpose, audience, and a range of formal and informal tasks.
>
> - SL.11-12.5. Make strategic use of digital media (e.g., textual, graphical, audio, visual, and interactive elements) in presentations to enhance understanding of findings, reasoning, and evidence and to add interest.
>
> - SL.11-12.6. Adapt speech to a variety of contexts and tasks, demonstrating a command of formal English when indicated or appropriate.

Continued

Table 4.4. (*continued*)

FRAMEWORK FOR 21ST CENTURY LEARNING
• Interdisciplinary Themes: Global Awareness; Financial, Economic, Business and Entrepreneurial Literacy; Civic Literacy; Environmental Literacy
• Learning and Innovation Skills: Creativity and Innovation, Critical Thinking and Problem Solving, Communication and Collaboration
• Information, Media, and Technology Skills: Information Literacy, Media Literacy, ICT Literacy
• Life and Career Skills: Flexibility and Adaptability, Initiative and Self-Direction, Social and Cross-Cultural Skills, Productivity and Accountability, Leadership and Responsibility

Table 4.5. Key Vocabulary for Lesson 2

Key Vocabulary	Definition
chain reaction	a series of reactions in which the product of the reaction causes additional reactions to take place
fertile material	a material that does not undergo fission reactions on its own but can be converted into a fissile material by irradiation in a reactor that results in neutron absorption; uranium-238 and thorium-232 are both fertile materials
fissile material	a material that can sustain nuclear fission chain reactions; the three primary fissile materials are uranium-233, uranium-235, and plutonium-239
fission	a process in which the nucleus of a heavy atom is split apart, either by an impact with another particle or by radioactive decay
nuclear reactor	a device used to initiate and control a sustained nuclear chain reaction, producing heat energy that is used to generate steam that drives a turbine

TEACHER BACKGROUND INFORMATION
Nuclear Fission Reactions and Nuclear Reactors

To teach this lesson, you should understand how fission reactions occur, how fission reactions are used in nuclear reactors, and how to calculate the general energy yields of fission reactions. You should also understand how to use exponential models to represent the energy release of a fission reaction.

Nuclear fission is a process in which the nucleus of a heavy atom is split apart, either by an impact with another particle or by radioactive decay. Uranium-235 is the fissile material most commonly used in nuclear reactors. Uranium is one of the heaviest elements; it is present in most rocks and is also found in ocean water. The uranium in Earth's crust is mostly U-238, an isotope that decays very slowly, with a half-life of about 4.5 billion years. U-235 decays somewhat faster than U-238 and is therefore more radioactive. U-235 is the only naturally occurring isotope that is fissile, although it makes up less than 1% of all naturally occurring uranium. (Uranium-233, plutonium-239, and plutonium-241 are also fissile materials used in reactors, but they are produced artificially using fertile isotopes such as thorium-232, uranium-238, and plutonium-240.)

The nucleus of U-235 contains 92 protons and 143 neutrons. In a nuclear reactor, the U-235 fission reaction is initiated by adding alpha-particle-emitting elements that release neutrons (see Figure 4.7 on p. 97). The U-235 captures the free neutrons, which split the nucleus in two (the resulting fragments may each be one of a number of elements found in the middle of the periodic table) and causes two or three additional neutrons to be emitted. These neutrons go on to cause additional fission reactions, causing the nuclei of more U-235 to split. This results in a chain reaction that can be sustained as long as there are U-235 nuclei present. The ongoing chain reaction causes a large amount of heat to be released. Most of the heat released (about 85%) comes from the kinetic energy of the fission fragments, while a smaller amount comes from the gamma rays emitted and the kinetic energy of emitted neutrons. This heat is used to create steam, which turns a turbine and drives a generator that produces electricity. Additional heat is produced in the reactor core from the radioactive decay of fission products. This heat will continue to be produced even after the chain reaction is stopped. This occurred in the Fukushima accident, when the core continued to produce heat after reactor shutdown.

Fission reactions can release large amounts of energy. When U-235 undergoes fission, it splits into two or more elements, such as $^{235}_{92}U + ^{1}_{0}n \rightarrow ^{142}_{56}Ba + ^{91}_{36}Kr + 3^{1}_{0}n$. Comparing the masses of the elements of the end product with the mass of the original element, it becomes evident that there is a net loss in mass. During the fission reaction, kinetic energy from the fission products is released, causing a measurable decrease in weight of nuclear particles. The energy released can be calculated by using Einstein's massenergy equivalence equation, $E = mc^2$, where E = energy, m = mass, and c = speed of light. Although the masses in question may be small, the energy yield can be enormous, since the mass is multiplied by the speed of light (186,000 miles/second or 300,000 km/second) squared. You may wish to remind students of the law of conservation of energy and point out that this means energy is not created during a fission reaction; instead, the energy of nuclear bonds is released.

Since each fission reaction creates large amounts of energy, the chain reaction must be controlled. This is accomplished by using control rods and moderator substances. The rods, made of such elements as silver, boron, or hafnium, absorb neutrons, thereby slowing the chain reaction, and can be inserted or withdrawn to regulate the amount of heat generated. The fuel in the reactor is surrounded by a moderator substance that slows the movement of the emitted neutrons and therefore slows the chain reaction. Moderator substances include water, heavy water, and graphite, depending on the type of reactor.

Nuclear reactors are frequently classified according to the type of coolant used. Several types of nuclear reactors are in operation worldwide, including pressurized water reactors (PWRs) (see Figure 4.8 on p. 98), boiling water reactors (BWRs), liquid metal–cooled reactors, gas-cooled reactors, and molten salt reactors. The majority of the nuclear reactors in operation in the United States as of this writing are PWRs, and therefore the focus of this module is on this type of reactor.

More information on nuclear fission reactions and nuclear reactors can be found on the following websites (the internet resources listed in the Teacher Background Information section in Lesson 1 on p. 52 may also be helpful):

- U.S. Energy Information Administration, *www.eia.gov/energyexplained/index.cfm?page=nuclear_home*

- World Nuclear Association, *www.world-nuclear.org/information-library/nuclear-fuel-cycle/introduction/what-is-uranium-how-does-it-work.aspx*

- U.S. Nuclear Regulatory Commission, *www.nrc.gov/reading-rm/basic-ref/students/for-educators.html* and *www.nrc.gov/reactors/pwrs.html*

- Virtual Nuclear Tourist, *www.nucleartourist.com*

- National Museum of Nuclear Science and History, *www.nuclearmuseum.org/learn/for-teachers/teacher-resources*

COMMON MISCONCEPTIONS

Students will have various types of prior knowledge about the concepts introduced in this lesson. Table 4.6 outlines some common misconceptions students may have concerning these concepts. Because of the breadth of students' experiences, it is not possible to anticipate every misconception that students may bring as they approach this lesson. Incorrect or inaccurate prior understanding of concepts can influence student learning in the future, however, so it is important to be alert to misconceptions such as those presented in the table.

Table 4.6. Common Misconceptions About the Concepts in Lesson 2

Topic	Student Misconception	Explanation
Nuclear energy	Nuclear reactions in nuclear power plants produce electricity.	In nuclear power plants, nuclear fission reactions produce energy that is used to heat water to turn a turbine that produces electricity.
	Nuclear power is too expensive to use in the United States.	In 2017, there were 61 nuclear power plants operating in 30 states, and they provided about 20% of the electricity in the United States.
	Nuclear power plants release dangerous amounts of radioactive material into the environment.	While nuclear power plants release small amounts of radioactive energy, the amounts they release in regular operation are far below legal safety limits and account for a very small proportion of the radiation humans are exposed to. Most human exposure to radiation comes from sources such as cosmic rays, naturally occurring uranium in Earth's crust, naturally occurring radon gas, medical procedures such as X-rays, and consumer products.

PREPARATION FOR LESSON 2

Review the Teacher Background Information section (p. 86), assemble the materials for the lesson, and preview the videos recommended in the Learning Components section that follows. Be prepared to guide a discussion or debate on the pros and cons of using fission reactors as energy sources.

LEARNING COMPONENTS
Introductory Activity/Engagement

Connection to the Challenge: Begin each day of this lesson by directing students' attention to the driving question for the module and challenge: How does the use of nuclear energy to meet our energy demands affect society? Hold a brief discussion of how students' learning in the previous days' lessons contributed to their ability to create a response to the Gammatown crisis. You may wish to hold a class discussion, creating a class list of key ideas on chart paper, or have students create a STEM Research Notebook entry with this information.

Science Class: Using the timeline they created in Lesson 1, ask students to identify key events that contributed to humans' ability to harness the power of radioactivity. Focus students' attention on what they learned about the Manhattan Project, and hold a class discussion about how this weapons development project led to the current use of nuclear power for peaceful purposes. Emphasize that the Manhattan Project transformed nuclear fission from a theoretical possibility into a feasible technology and that this technology is used in today's nuclear reactors. Introduce the concept of nuclear fission using the U-235 fission reaction graphic (Figure 4.7 on p. 97). Ask students to brainstorm ideas about how this reaction releases energy.

Mathematics Class: Tell students that they are going to consider the amount of energy released by the first atomic bomb used in warfare. This bomb, given the code name "Little Boy," was dropped on Hiroshima, Japan, on August 6, 1945. Tell students that this bomb exploded with the energy of about 15 kilotons of TNT. Have students convert this amount into joules (J) and into kilowatt hours (kWh). You may have students research conversion factors or provide them with the following:

- 1 ton of TNT = 4,184,000,000 J

- 1 J = 0.0000002778 kWh

Students should express quantities using scientific notation.

After students have completed their calculations, have them conduct research to find out how much energy (in kWh) a typical home uses in one month and compare that with the energy released by "Little Boy."

ELA Connection: Have students read and analyze Albert Einstein's letter to President Franklin D. Roosevelt about the construction of nuclear bombs. A transcript can be found at *www.atomicarchive.com/Docs/Begin/Einstein.shtml.* Also see *www.fdrlibrary.marist. edu/archives/pdfs/docsworldwar.pdf* for an image of the original letter.

STEM Research Notebook Prompt

Students should respond to the following prompt in their STEM Research Notebooks, supporting their answers with evidence from Einstein's letter: *Did Albert Einstein support or oppose the creation of nuclear weapons?*

Social Studies Connection: Have the class explore the factors leading to the development of the first nuclear weapon, using a video that provides historical context for the development of nuclear weapons such as "A Brief History of the Atomic Bomb" at *www. youtube.com/watch?v=AZsngBRP8o0.* After watching this video, show Lyndon B. Johnson's famous 1964 campaign advertisement at *www.youtube.com/watch?v=2cwqHB6QeUw.* Hold a class discussion on students' perceptions of the changes in messages about atomic military power during the years between the end of World War II and Johnson's campaign ad.

Activity/Exploration

Science Class: Introduce fission reactions, emphasizing that energy is released in these reactions when nuclear bonds of heavy elements are broken. When this occurs, fragments with lower masses are released along with large amounts of energy, including kinetic energy of particles that are released in the fission reaction and energy that is released from decaying fission products. In this lesson, students create a diagram of a nuclear reactor and a model of nuclear fission. Students should use the EDP for both tasks.

Nuclear Reactor Diagram

Have student teams conduct research to understand the design and operation of a PWR. Each student should create a schematic diagram of a PWR, labeling each component. Students should understand that schematic drawings are not intended to be realistic drawings but rather use graphic symbols to represent the parts of a system and how they are interconnected. Emphasize that all parts should be labeled and that the operation of the PWR should be evident from their drawings. You may wish to direct students to the Nuclear Regulatory Commission's overview of the operation of a PWR, along with a graphic labeling the various parts of the reactor, at *www.nrc.gov/reactors/pwrs.html*.

Remind students of the steps of the EDP. You may wish to suggest that students use a jigsaw approach in their groups, with each student taking responsibility for investigating one component of reactor design, reporting back to the group on that component.

Emphasize to students that collaboration among team members is essential in solving problems as a group. Point out that it will be important to work well as a group to solve the Gammatown Crisis Challenge. Ask students to share what went well with their collaboration in Lesson 1 and what challenges they faced in collaborating to create their decay chain models. Ask them to provide guidelines for good collaboration, and create a class list of collaboration practices. A collaboration rubric is provided at the end of this lesson (pp. 101–102); you may wish to use this to guide the class discussion and provide a copy of the rubric to students.

Nuclear Fission Model

Each team should also create a model of nuclear fission as it occurs within a PWR. This model should focus on the U-235 fission reaction, using the reaction depicted in Figure 4.7 (p. 97) as a starting point. The models should include a neutron source, evidence of the chain reaction, and a scaled representation of the energy released in the reaction. Teams can use a variety of forms to create their models; for example, a model might be a computer simulation or a two-dimensional model. The energy component of the model will be supported by the mathematics activity described below.

Remind students to employ the steps of the EDP in creating their models. You may wish to have students create entries in their STEM Research Notebooks outlining how they used each step in the EDP to create their models.

Mathematics Class: Have students calculate the energy output of a single fission reaction using the reaction $^{235}_{92}U + ^{1}_{0}n \rightarrow ^{142}_{56}Ba + ^{91}_{36}Kr + 3^{1}_{0}n$. Students should use the following equation as a starting point: $\Delta BE = BE_{products} - BE_{reactants}$ (BE = binding energy). They should then compound the energy yield created in a chain reaction and show their findings in a table and graph. Students should use these findings for the Nuclear Fission Models they are creating in science class.

ELA Connection: Present students with the following scenario and have them respond to it on notebook paper or in their STEM Research Notebooks: *You have been transported in time to August 2, 1939. Because of your outstanding scientific knowledge, you have been given the opportunity to have a letter delivered to President Franklin D. Roosevelt to express your views on the development of nuclear weapons. You may draw on your knowledge of history to inform your views, but you cannot indicate to the president that you have any knowledge of events that have not yet happened. Write your own letter, beginning with the opening line from Einstein's letter to the president: "Some recent work by E. Fermi and L. Szilard, which has been communicated to me in manuscript, leads me to expect that the element uranium may be turned into a new and important source of energy in the immediate future."*

Social Studies Connection: Have students research how nuclear energy influenced U.S. foreign affairs and public perception of nuclear energy from the end of World War II through the present. You may wish to divide students into groups to research events or topics and have each group prepare a presentation for the class. Events might include the following:

- The Cold War

- The Cuban Missile Crisis

- The Strategic Defense Initiative

- U.S. relations with North Korea

- The Treaty on the Non-Proliferation of Nuclear Weapons

- The Three Mile Island nuclear accident

- The Chernobyl nuclear accident

- The Fukushima nuclear accident

Explanation

Science Class: Explain to students that the element thorium can be used in nuclear reactors instead of uranium. Ask students to share their ideas about why alternatives to uranium might be attractive (e.g., it may be difficult to access uranium, the cost of uranium). Have students locate the element thorium on the periodic table and compare it with uranium (e.g., different locations on periodic table, different atomic weights). Tell students that later in the lesson, they will research the use of thorium in nuclear reactors.

Mathematics Class: Review the use of exponential functions and summation formulas to efficiently calculate energy yields of fission reactions. Review scientific notation and unit conversions with students.

ELA Connection: Review business letter formats with students, using their letters to President Roosevelt as an example. Discuss the differences between primary and secondary sources, and have students provide examples of each and give advantages and disadvantages of using each type of source in research. Create a class chart with students' responses.

Social Studies Connection: Review internet search practices. You may wish to discuss varying the wording of search terms, conducting multiple searches, and using Boolean operators (AND, OR, and NOT) to make searches more specific. In addition, students should be able to identify reliable and credible sources of information and distinguish between peer-reviewed articles and articles in the popular press.

Elaboration/Application of Knowledge

Science Class: Have student teams research thorium fission reactors, their benefits as an alternative to traditional uranium fission reactors, and the means necessary to implement such technology on a large scale. Have each team create a jigsaw essay on their findings. To create this essay, each team member should research and write one component of the essay, such as the following:

- Availability and characteristics of thorium

- How thorium can be used in nuclear reactions in power plants

- Arguments for thorium reactors

- Challenges to using thorium reactors

The team members should then synthesize their written findings into one coherent essay. Resources about thorium reactors include the following:

- *https://en.wikipedia.org/wiki/Thorium_fuel_cycle*

- *www.ted.com/talks/taylor_wilson_my_radical_plan_for_small_nuclear_fission_reactors*

- *www.youtube.com/watch?v=IZf6e0ntFrw*

- *www.youtube.com/watch?v=N2vzotsvvkw*

- *www.thoriumenergyworld.com*

To review and extend their understanding of nuclear energy, have students play the Fermi Feud quiz game from the U.S. Department of Energy at *www.energy.gov/ne/downloads/fermi-feud*, discussing the concepts and new vocabulary terms throughout the game.

STEM Research Notebook Prompt

Have students respond to this prompt in their STEM Research Notebooks: *Based on what you have learned so far about nuclear energy, what is your opinion about the safety of nuclear power plants? Be sure to provide evidence for your assertions, and conclude by answering the question, Are nuclear fission reactions a good alternative to coal and natural gas for fulfilling our nation's energy needs?*

Mathematics Class: Working as a class, use students' tabular and graphic data from their chain reaction models to determine an exponential regression model equation that allows them to predict energy yields for a given mass of U-235. Then, have students investigate the yearly energy needs of their city or town and calculate how much U-235 would be required to meet these needs.

ELA Connection: Have student teams create a radio public service announcement that either supports or opposes constructing a nuclear power plant in their region.

Social Studies Connection: Have students investigate the health effects of radioactivity, focusing on the biological effects on humans and evidence of these effects from documented cases of radiation exposure.

Then, have students explore hindrances to the development of nuclear energy, paying special attention to the ecological ramifications of its use and the anti-nuclear movement. Divide students into two teams (one that is for and one that is against nuclear energy) and hold a debate about whether the United States should focus on nuclear energy to provide future energy needs.

Evaluation/Assessment

Students may be assessed on the following performance tasks and other measures listed.

Performance Tasks

- Nuclear reactor schematic diagram

- Nuclear Fission Model Rubric (p. 99)

- Thorium Reactor Jigsaw Essay Rubric (p. 100)

- Letter to President Roosevelt

- Team nuclear history presentations

Other Measures

- STEM Research Notebook entries

- Collaboration Rubric (pp. 101–102)

INTERNET RESOURCES

Nuclear energy, fission reactions, and nuclear reactors

- *www.eia.gov/energyexplained/index.cfm?page=nuclear_home*

- *www.world-nuclear.org/information-library/nuclear-fuel-cycle/introduction/ what-is-uranium-how-does-it-work.aspx*

- *www.nrc.gov/reading-rm/basic-ref/students/for-educators.html*

- *www.nrc.gov/reactors/pwrs.html*

- *www.nucleartourist.com*

- *www.nuclearmuseum.org/learn/for-teachers/teacher-resources*

Einstein's letter to President Franklin D. Roosevelt

- *www.atomicarchive.com/Docs/Begin/Einstein.shtml*

- *www.fdrlibrary.marist.edu/archives/pdfs/docsworldwar.pdf*

"A Brief History of the Atomic Bomb" video

- *www.youtube.com/watch?v=AZsngBRP8o0*

Lyndon B. Johnson's 1964 campaign ad video

- *www.youtube.com/watch?v=2cwqHB6QeUw*

Thorium reactors

- *https://en.wikipedia.org/wiki/Thorium_fuel_cycle*

- *www.ted.com/talks/taylor_wilson_my_radical_plan_for_small_nuclear_fission_reactors*

- *www.youtube.com/watch?v=IZf6e0ntFrw*

- *www.youtube.com/watch?v=N2vzotsvvkw*

- *www.thoriumenergyworld.com*

Fermi Feud game
- *www.energy.gov/ne/downloads/fermi-feud*

FIGURE 4.7. URANIUM-235 NUCLEAR FISSION REACTION

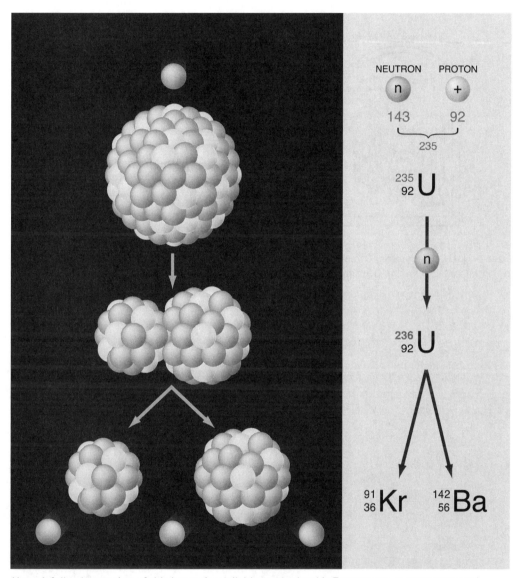

Note: A full-color version of this image is available on the book's Extras page at *www.nsta.org/roadmap-radioactivity.*

FIGURE 4.8. PRESSURIZED WATER REACTOR (PWR)

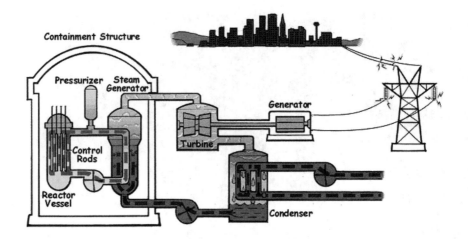

Source: Nuclear Regulatory Commission at *www.nrc.gov/reading-rm/basic-ref/students/animated-pwr.html* (public domain). (This image is animated on the website.)

Nuclear Fission Model Rubric

Team Name: _____

Type of Model	Needs Improvement (1 point)	Approaches Expectations (2 points)	Meets Expectations (3 points)	Exceeds Expectations (4 points)	Score
PHYSICAL MODEL	The model is missing major components and does not represent fission reactions correctly.	The model includes the correct initial element but is lacking products or energy emissions.	The model is created from 3-D objects and includes the reactants, products, neutrons, and energy emissions associated with the reactions, but it may contain a few minor errors.	The model is created from 3-D objects and includes the reactants, products, neutrons, and energy emissions associated with the reactions.	
COMPUTER-GENERATED MODEL	The model is a single computer-generated picture of the fission process. There may be several mistakes.	The model demonstrates the fission process through computer-generated images or frames that are essentially pictures. There are mistakes in the model.	The model aptly demonstrates the fission process through some form of stop-motion animation (not fluid) that shows neutron bombardment, reactants, products, and energy emissions. The initial model is accurate with few flaws.	The model aptly demonstrates the fission process through some form of fluid animation that shows neutron bombardment, reactants, products, and energy emissions. The initial model is accurate and without flaws.	
2-D MODEL	The model is inaccurate, and there are errors in the depiction of the fission process and energy emissions.	While the model may be created well, there are errors in the depiction of the fission process or energy emissions.	The model is created well, although there may be minor errors in its depiction of the fission process.	Model is hand-drawn or created with computer software. It is well organized and includes the reactants, products, neutrons, and energy emissions associated with the reactions.	

TOTAL SCORE: _____

COMMENTS:

Thorium Reactor Jigsaw Essay Rubric

Team Name: _____

Student Name: _____

Criteria	Needs Improvement (1 point)	Approaches Expectations (2 points)	Meets Expectations (3 points)	Exceeds Expectations (4 points)	Score
ORGANIZATION OF ESSAY	One or more sections are missing, and it is clear that the research team did not work together to weave their sections together.	The individual sections of the essay were not linked sufficiently as to make the essay flow well and make sense to the reader.	The essay is coherent and flows well; the sections connect decently with a few errors.	Students worked together to create a clear, coherent, and cogent essay describing all criteria of the performance task.	
INDIVIDUAL CONTRIBUTION	Student's research does not outline the main tenets of his/her section; many important aspects of the section are missing.	Student's research outlines most of the key components of his/her section, but some are missing.	Student research outlines the key components of his/her section, but some components are not fully described.	Student research is thorough and clearly presented in his/her section of the essay.	
INDIVIDUAL WRITING QUALITY	The student's writing is very difficult to follow; the information for his/her section is not adequately described.	The student's writing is sufficient for understanding his/her section of the essay; there are several errors in grammar and syntax.	The student's writing is easy to follow, with few errors in grammar or syntax.	The student's writing is clear and concise, with few errors in grammar and syntax.	
TEAMWORK	The essay is a scattering of individual work with little to no evidence of teamwork.	Students worked together well to outline their paper but did not spend sufficient time weaving individual sections together into a coherent essay.	Students worked well together to outline their essay and wove their individual sections together decently; one or two sections needed clearer segues.	Students clearly worked well together to outline their essay and disseminated their individual research to one another to create a clear, coherent, and cogent product.	

TOTAL SCORE: _____

COMMENTS:

Continued

Collaboration Rubric

Student Name: _____ Team Name: _____

Individual Performance	Needs Improvement (0–3 points)	Approaches Expectations (4–7 points)	Meets or Exceeds Expectations (9–10 points)	Score
INDIVIDUAL ACCOUNTABILITY	• Student is unprepared. • Student does not communicate with team members and does not manage tasks as agreed on by the team. • Student does not complete or participate in project tasks. • Student does not complete tasks on time. • Student does not use feedback from others to improve work.	• Student is usually prepared. • Student sometimes communicates with team members and manages tasks as agreed on by the team, but not consistently. • Student completes or participates in some project tasks but needs to be reminded. • Student completes most tasks on time. • Student sometime uses feedback from others to improve work.	• Student is consistently prepared. • Student consistently communicates with team members and manage tasks as agreed on by the team. Stucent discusses and reflects on ideas with the team. • Student completes or participates in project tasks without being reminded. • Student completes tasks on time. • Student uses feedback from others to improve work.	
TEAM PARTICIPATION	• Student does not help the team solve problems; may interfere with team work. • Student does not express ideas clearly, pose relevant questions, or participate in group discussions. • Student does not give useful feedback to other team members. • Student does not volunteer to help others when needed.	• Student cooperates with the team but may not actively help solve problems. • Student sometimes expresses ideas, poses relevant questions, elaborates in response to questions, and participates in group discussions. • Student provides some feedback to team members. • Student sometimes volunteers to help others.	• Student helps the team solve problems and manage conflicts. • Student makes discussions effective by clearly expressing ideas, posing questions, and responding thoughtfully to team members' questions and perspectives. • Student gives useful feedback to others so they can improve their work. • Student volunteers to help others if needed.	

Collaboration Rubric (*continued*)

Individual Performance	Needs Improvement (0–3 points)	Approaches Expectations (4–7 points)	Meets or Exceeds Expectations (9–10 points)	*Score*
PROFESSIONALISM AND RESPECT FOR TEAM MEMBERS	• Student is impolite or disrespectful to other team members. • Student does now acknowledge or respect others' ideas and perspectives.	• Student is usually polite and respectful to other team members. • Student usually acknowledges and respects others' ideas and perspectives.	• Student is consistently polite and respectful to other team members. • Student consistently acknowledges and respects others' ideas and perspectives.	

TOTAL SCORE: _____

COMMENTS:

Lesson Plan 3: Nuclear Fusion: Harnessing the Power of the Stars

In this lesson, students consider nuclear fusion and its potential to meet human energy needs in the future, as well as the substantial technology and engineering challenges associated with creating fusion reactions. Students compare and contrast the processes of fission and fusion and the associated benefits and concerns about using these two processes as energy sources. *Note:* If there is insufficient time to complete all lessons in the module, this lesson may be omitted or taught at a later time, after the completion of the module challenge.

ESSENTIAL QUESTIONS

- What is nuclear fusion?

- Why is nuclear fusion so difficult to generate?

- Would using nuclear fusion in power plants be safer and cleaner for the environment than using nuclear fission?

ESTABLISHED GOALS AND OBJECTIVES

At the conclusion of this lesson, students will be able to do the following:

- Explain the differences between fission and fusion reactions

- Explain the potential of nuclear fission as a power source

- Describe the challenges associated with using nuclear fission as a power source

- Create a model of nuclear fusion, exhibiting an understanding of why intense heat and pressure are necessary to generate a fusion reaction

- Use their understanding of fusion energy to develop a novel and useful product

- Use their understanding of fusion to clearly explain to others the product developed by student's team in challenge

- Apply the EDP to solve a complex problem

- Collaborate with peers to solve a problem

TIME REQUIRED

- 4 days (approximately 45 minutes each; see Table 3.8, pp. 40)

MATERIALS

Required Materials for Lesson 3

- STEM Research Notebooks

- Computers with internet access for student research and viewing videos (for each team)

- Periodic tables (1 per student)

- Software for 3-D modeling (for teams who choose to create computer-simulated models of nuclear fusion)

- Art supplies (for teams who choose to make physical models of nuclear fusion)

- Indirectly vented chemical splash safety goggles

- Handouts (attached at the end of this lesson)

Additional Materials for Product Development Challenge (per team)

- Option A: Poster or presentation board, set of markers, 5 or 6 sheets of colored paper, glue, and scissors

- Option B: Presentation software

SAFETY NOTES

1. All students must wear safety goggles during all phases of this inquiry activity.

2. Use caution when working with sharps (scissors) to avoid cutting or puncturing skin.

3. Wash hands with soap and water after the activity is completed.

CONTENT STANDARDS AND KEY VOCABULARY

Table 4.7 lists the content standards from the *NGSS, CCSS,* and the Framework for 21st Century Learning that this lesson addresses, and Table 4.8 (p. 111) presents the key vocabulary. Vocabulary terms are provided for both teacher and student use. Teachers may choose to introduce some or all of the terms to students.

Table 4.7. Content Standards Addressed in STEM Road Map Module Lesson 3

NEXT GENERATION SCIENCE STANDARDS

PERFORMANCE EXPECTATIONS

- HS-PS1-1. Use the periodic table as a model to predict the relative properties of elements based on the patterns of electrons in the outermost energy level of atoms.

- HS-PS1-7. Use mathematical representations to support the claim that atoms, and therefore mass, are conserved during a chemical reaction.

- HS-PS1-8. Develop models to illustrate the changes in the composition of the nucleus of the atom and the energy released during the processes of fission, fusion, and radioactive decay.

- HS-ETS-2. Design a solution to a complex real-world problem by breaking it down into smaller, more manageable problems that can be solved through engineering.

SCIENCE AND ENGINEERING PRACTICES

Using Mathematics and Computational Thinking

- Use mathematical representations of phenomena to support claims.

- Create a computational model or simulation of a phenomenon, designed device, process, or system.

Developing and Using Models

- Use a model to predict the relationships between systems or between components of a system.

- Develop a model based on evidence to illustrate the relationships between systems or between components of a system.

Obtaining and Communicating Information

- Evaluate the validity and reliability of multiple claims that appear in scientific and technical texts or media reports, verifying the data when possible.

- Communicate technical information or ideas in multiple formats.

Constructing Explanations and Designing Solutions

- Design, evaluate, and/or refine a solution to a complex real-world problem, based on scientific knowledge, student-generated sources of evidence, prioritized criteria, and trade-off considerations.

- Construct and revise an explanation based on valid and reliable evidence obtained from a variety of sources (including students' own investigations, models, theories, simulations, peer review) and the assumption that theories and laws that describe the natural world operate today as they did in the past and will continue to do so in the future.

Continued

Table 4.7. (*continued*)

DISCIPLINARY CORE IDEAS

PS1.A: Structure and Properties of Matter

- Each atom has a charged substructure consisting of a nucleus, which is made of protons and neutrons, surrounded by electrons.

- The periodic table orders elements horizontally by the number of protons in the atom's nucleus and places those with similar chemical properties in columns. The repeating patterns of this table reflect patterns of outer electron states.

- A stable molecule has less energy than the same set of atoms separated; one must provide at least this energy in order to take the molecule apart.

PS1.B: Chemical Reactions

- The fact that atoms are conserved, together with knowledge of the chemical properties of the elements involved, can be used to describe and predict chemical reactions.

PS1.C: Nuclear Processes

- Nuclear processes, including fusion, fission, and radioactive decays of unstable nuclei, involve release or absorption of energy. The total number of neutrons plus protons does not change in any nuclear process.

- Spontaneous radioactive decays follow a characteristic exponential decay law. Nuclear lifetimes allow radiometric dating to be used to determine the ages of rocks and other materials.

PS2.B: Types of Interactions

- Attraction and repulsion between electric charges at the atomic scale explain the structure, properties, and transformations of matter, as well as the contact forces between material objects.

PS3.A: Definitions of Energy

- Energy is a quantitative property of a system that depends on the motion and interactions of matter and radiation within that system. That there is a single quantity called energy is due to the fact that a system's total energy is conserved, even as, within the system, energy is continually transferred from one object to another and between its various possible forms.

PS3.B: Conservation of Energy and Energy Transfer

- Conservation of energy means that the total change of energy in any system is always equal to the total energy transferred into or out of the system.

- Energy cannot be created or destroyed, but it can be transported from one place to another and transferred between systems.

Continued

Table 4.7. (*continued*)

DISCIPLINARY CORE IDEAS (*continued*)

PS3.B: Conservation of Energy and Energy Transfer (*continued*)

- Mathematical expressions, which quantify how the stored energy in a system depends on its configuration (e.g., relative positions of charged particles, compression of a spring) and how kinetic energy depends on mass and speed, allow the concept of conservation of energy to be used to predict and describe system behavior.

- The availability of energy limits what can occur in any system.

PS3.D: Energy in Chemical Processes

- Although energy cannot be destroyed, it can be converted to less useful forms—for example, to thermal energy in the surrounding environment.

- Nuclear fusion processes in the center of the sun release the energy that ultimately reaches Earth as radiation.

ETS1.A: Defining and Delimiting an Engineering Problem

- Criteria and constraints also include satisfying any requirements set by society, such as taking issues of risk mitigation into account, and they should be quantified to the extent possible and stated in such a way that one can tell if a given design meets them.

ETS1.C: Optimizing the Design Solution

- Criteria may need to be broken down into simpler ones that can be approached systematically, and decisions about the priority of certain criteria over others (trade-offs) may be needed.

CROSSCUTTING CONCEPTS

Patterns

- Different patterns may be observed at each of the scales at which a system is studied and can provide evidence for causality in explanations of phenomena.

Energy and Matter

- Changes of energy and matter in a system can be described in terms of energy and matter flows into, out of, and within that system.

- In nuclear processes, atoms are not conserved, but the total number of protons plus neutrons is conserved.

- The total amount of energy and matter in closed systems is conserved.

Stability and Change

- Much of science deals with constructing explanations of how things change and how they remain stable.

Continued

Table 4.7. (*continued*)

CROSSCUTTING CONCEPTS (*continued*)

Cause and Effect

- Cause and effect relationships can be suggested and predicted for complex natural and human designed systems by examining what is known about smaller scale mechanisms within the system.

Scale, Proportion, and Quantity

- The significance of a phenomenon is dependent on the scale, proportion, and quantity at which it occurs.

Systems and System Models

- Models can be used to predict the behavior of a system, but these predictions have limited precision and reliability due to the assumptions and approximations inherent in models.

Structure and Function

- Investigating or designing new systems or structures requires a detailed examination of the properties of different materials, the structures of different components, and connections of components to reveal its function and/or solve a problem.

COMMON CORE STATE STANDARDS FOR MATHEMATICS

MATHEMATICAL PRACTICES

- MP1. Make sense of problems and persevere in solving them.
- MP3. Construct viable arguments and critique the reasoning of others.
- MP4. Model with mathematics.
- MP7. Look for and make use of structure.

MATHEMATICAL CONTENT

- HSA.APR.D.6. Rewrite simple rational expressions in different forms; write $a(x)/b(x)$ in the form $q(x) + r(x)/b(x)$, where $a(x)$, $b(x)$, $q(x)$, and $r(x)$ are polynomials with the degree of $r(x)$ less than the degree of $b(x)$ using inspection, long division, or, for the more complicated examples, a computer algebra system.
- HSF.IF.B.4. For a function that models a relationship between two quantities, interpret key features of graphs and tables in terms of the quantities, and sketch graphs showing key features given a verbal description of the relationship.
- HSF.IF.B.6. Calculate and interpret the average rate of change of a function (presented symbolically or as a table) over a specified interval. Estimate the rate of change from a graph.
- HSF.LE.A.1. Distinguish between situations that can be modeled with linear functions and with exponential functions.

Continued

Table 4.7. (*continued*)

> ## MATHEMATICAL CONTENT (*continued*)
>
> - HSF.LE.A.2. Construct linear and exponential functions, including arithmetic and geometric sequences, given a graph, a description of a relationship, or two input-output pairs (include reading these from a table).
>
> - HSF.LE.A.3. Observe using graphs and tables that a quantity increasing exponentially eventually exceeds a quantity increasing linearly, quadratically, or (more generally) as a polynomial function.
>
> - HSF.LE.A.4. For exponential models, express as a logarithm the solution to $ab^{ct} = d$ where a, c, and d are numbers and the base b is 2, 10, or e; evaluate the logarithm using technology.
>
> - HSF.LE.B.5. Interpret the parameters in a linear or exponential function in terms of a context.
>
> - HSF.BF.A.1. Write a function that describes a relationship between two quantities.
>
> - HSF.BF.A.2. Write arithmetic and geometric sequences both recursively and with an explicit formula, use them to model situations, and translate between the two forms.
>
> - HSF.BF.B.3. Identify the effect on the graph of replacing $f(x)$ by $f(x) + k$, $k\,f(x)$, $f(kx)$, and $f(x + k)$ for specific values of k (both positive and negative); find the value of k given the graphs. Experiment with cases and illustrate an explanation of the effects on the graph using technology. Include recognizing even and odd functions from their graphs and algebraic expressions for them.
>
> - HSF.BF.B.4. Find inverse functions.
>
> - HSF.BF.B.5. Understand the inverse relationship between exponents and logarithms and use this relationship to solve problems involving logarithms and exponents.
>
> ## COMMON CORE STATE STANDARDS FOR ENGLISH LANGUAGE ARTS
>
> ### READING STANDARDS
>
> - RI.11-12.1. Cite strong and thorough textual evidence to support analysis of what the text says explicitly as well as inferences drawn from the text, including determining where the text leaves matters uncertain.
>
> - RI.11-12.2. Determine two or more central ideas of a text and analyze their development over the course of the text, including how they interact and build on one another to provide a complex analysis; provide an objective summary of the text.
>
> - RI.11-12.3. Analyze a complex set of ideas or sequence of events and explain how specific individuals, ideas, or events interact and develop over the course of the text.
>
> - RI.11-12.4. Determine the meaning of words and phrases as they are used in a text, including figurative, connotative, and technical meanings; analyze how an author uses and refines the meaning of a key term or terms over the course of a text.

Continued

Table 4.7. (*continued*)

READING STANDARDS (*continued*)

- RI.11-12.5. Analyze and evaluate the effectiveness of the structure an author uses in his or her exposition or argument, including whether the structure makes points clear, convincing, and engaging.

- RI.11-12.7. Integrate and evaluate multiple sources of information presented in different media or formats (e.g., visually, quantitatively) as well as in words in order to address a question or solve a problem.

- RI.11-12.8. Delineate and evaluate the reasoning in seminal U.S. texts, including the application of constitutional principles and use of legal reasoning and the premises, purposes, and arguments in works of public advocacy.

WRITING STANDARDS

- W.11-12.1. Write arguments to support claims in an analysis of substantive topics or texts, using valid reasoning and relevant and sufficient evidence.

- W.11-12.2. Write informative/explanatory texts to examine and convey complex ideas, concepts, and information clearly and accurately through the effective selection, organization, and analysis of content.

SPEAKING AND LISTENING STANDARDS

- SL.11-12.1. Initiate and participate effectively in a range of collaborative discussions (one-on-one, in groups, and teacher-led) with diverse partners on grades 11–12 topics, texts, and issues, building on others' ideas and expressing their own clearly and persuasively.

- SL.11-12.2. Integrate multiple sources of information presented in diverse formats and media (e.g., visually, quantitatively, orally) in order to make informed decisions and solve problems, evaluating the credibility and accuracy of each source and noting any discrepancies among the data.

- SL.11-12.3. Evaluate a speaker's point of view, reasoning, and use of evidence and rhetoric, assessing the stance, premises, links among ideas, word choice, points of emphasis, and tone used.

- SL.11-12.4. Present information, findings, and supporting evidence, conveying a clear and distinct perspective, such that listeners can follow the line of reasoning, alternative or opposing perspectives are addressed, and the organization, development, substance, and style are appropriate to purpose, audience, and a range of formal and informal tasks.

- SL.11-12.5. Make strategic use of digital media (e.g., textual, graphical, audio, visual, and interactive elements) in presentations to enhance understanding of findings, reasoning, and evidence and to add interest.

- SL.11-12.6. Adapt speech to a variety of contexts and tasks, demonstrating a command of formal English when indicated or appropriate.

Continued

Table 4.7. (*continued*)

> **FRAMEWORK FOR 21ST CENTURY LEARNING**
> - Interdisciplinary Themes: Global Awareness; Financial, Economic, Business and Entrepreneurial Literacy; Civic Literacy; Environmental Literacy
> - Learning and Innovation Skills: Creativity and Innovation, Critical Thinking and Problem Solving, Communication and Collaboration
> - Information, Media, and Technology Skills: Information Literacy, Media Literacy, ICT Literacy
> - Life and Career Skills: Flexibility and Adaptability, Initiative and Self-Direction, Social and Cross-Cultural Skills, Productivity and Accountability, Leadership and Responsibility

Table 4.8. Key Vocabulary for Lesson 3

Key Vocabulary	Definition
Coulomb barrier	the energy barrier that two nuclei must overcome to fuse in a nuclear reaction
deuterium	a stable isotope of hydrogen containing one proton and one neutron; known as heavy hydrogen
hydrogen isotopes	hydrogen atoms containing one proton and between zero and two neutrons; protium, deuterium, and tritium are the three naturally occurring hydrogen isotopes
nuclear force	the force that holds subatomic particles (protons and neutrons) together in the nucleus of an atom; this force is extremely strong but works only over extremely small distances
nuclear fusion	a nuclear reaction in which two or more atomic nuclei collide at high speeds and join, forming a new atomic nucleus and releasing large amounts of energy
supernova	the explosion of a large star (several times more massive than our Sun), occurring at the end of the star's lifetime when its hydrogen and helium fuel is depleted; this results in the star contracting in on itself until the core eventually becomes heavy enough that the star implodes and emits elements into space
tritium	a radioactive isotope of hydrogen with one proton and two neutrons; known as hydrogen-3

TEACHER BACKGROUND INFORMATION
Nuclear Fusion

For this lesson, you should be familiar with the process of nuclear fusion, its potential, and its limitations. The prospect of using nuclear fusion is appealing because it has the potential to provide an abundant source of energy for the future. As of 2019, however, the technology and engineering challenges associated with harnessing the power of nuclear fusion have been daunting and met with only limited success, and therefore fusion is not currently a feasible source of energy to meet human needs.

Nuclear fusion is the process in which two atomic nuclei fuse to form new atoms (see Figure 4.9 on p. 118). This process occurs in stars such as our Sun, where hydrogen atoms fuse to become helium atoms, and energy is released. Fusion is not possible under normal circumstances, since the positively charged nuclei of atoms repel one another, preventing the atoms from coming close enough to collide. In conditions where nuclei can overcome the electrostatic forces that cause them to repel one another (for instance, when the temperature increases enough that atoms move at very high speeds), the attractive forces of the nuclei (the forces that hold the nuclei together) permit them to fuse. One of the key barriers to achieving fusion reactions is the substantial amount of energy required to overcome what is known as the Coulomb barrier, the repellent electromagnetic forces that two nuclei must overcome to fuse. Particles must have an enormous amount of thermal energy to exceed the Coulomb barrier. The high temperatures of the Sun and other stars are sufficient for thermonuclear fusion; however, this type of fusion is not feasible on Earth.

Reactions involving deuterium-tritium fusion hold the most potential for harnessing nuclear fusion as an energy source on Earth. This reaction releases about four times the amount of energy as uranium fission, given equal masses of reactants. Deuterium is readily available, occurring naturally in ocean water. Tritium is available only in trace quantities in nature and has a half-life of just 12 years, but it can be made in nuclear reactors. The challenge has been to develop technology to heat the deuterium-tritium fuel to high enough temperatures and confine it for sufficient time so that more energy is released than is used to initiate the reaction. Current research focuses on magnetic and inertial confinement to achieve these conditions. The World Nuclear Organization provides an overview of fusion and current fusion technologies at *www.world-nuclear.org/ information-library/current-and-future-generation/nuclear-fusion-power.aspx*.

Common Misconceptions

Students will have various types of prior knowledge about the concepts introduced in this lesson. Table 4.9 outlines some common misconceptions students may have concerning these concepts. Because of the breadth of students' experiences, it is not possible to anticipate every misconception that students may bring as they approach this lesson.

Incorrect or inaccurate prior understanding of concepts can influence student learning in the future, however, so it is important to be alert to misconceptions such as those presented in the table.

Table 4.9. Common Misconceptions About the Concepts in Lesson 3

Topic	Student Misconception	Explanation
Nuclear science and technology	Fission and fusion reactions are the same, except that fusion is more powerful than fission.	In fission reactions, atoms are split, whereas in fusion reactions, atoms are combined.
	Fusion reactions do not involve any radioactivity.	While deuterium is not radioactive, tritium, another isotope of hydrogen, is radioactive and emits beta particles.
	Once humans achieve nuclear fusion, it will provide an ideal way to meet human needs for electricity.	Scientists have produced fusion reactions at a very small scale; however, fusion reactions consume far more energy than they produce.
	Nuclear fusion can provide an unlimited source of power for humans.	Because naturally occurring supplies of the tritium needed for fusion reactions are scarce, tritium must be produced by irradiating lithium. Thus, the potential for human use of nuclear fusion is limited by supplies of lithium.

PREPARATION FOR LESSON 3

Review the Teacher Background Information section (p. 112), assemble the materials for the lesson, make copies of the student handout, and preview the videos recommended in the Learning Components section that follows.

LEARNING COMPONENTS
Introductory Activity/Engagement

Connection to the Challenge: Begin each day of this lesson by directing students' attention to the driving question for the module and challenge: How does the use of nuclear energy to meet our energy demands affect society? Hold a brief discussion of how students' learning in the previous days' lessons contributed to their ability to create a response to the Gammatown crisis. You may wish to hold a class discussion, creating a class list of key ideas on chart paper, or have students create a STEM Research Notebook entry with this information.

Science Class: Show students a video about nuclear fusion and issues surrounding its practical applications, such as "What Is Fusion?" at *www.phdcomics.com/comics.php?f=1716.*

Hold a class discussion about fusion, posing questions such as the following:

- How is nuclear fusion different from nuclear fission?

- Based on what you saw in the video, do you think that nuclear fusion is a feasible alternative energy source? Why or why not?

- What are some of the benefits of fusion energy versus other forms of energy?

- What are the challenges of fusion energy?

- What is the "strong force" the narrators refer to?

Mathematics Class: Students need to be familiar with the atomic masses of deuterium and tritium for this lesson. Provide students with periodic tables and the information that deuterium is hydrogen with an extra neutron and tritium is hydrogen with two neutrons. A neutron has the atomic mass of 1.0086654 atomic mass units (amu). Challenge students to calculate the atomic mass of deuterium and tritium in amu and convert this to grams, expressing the mass in scientific notation. Students should use the conversion factor 1 amu = 1.66×10^{-24} grams.

ELA Connection: Not applicable.

Social Studies Connection: Not applicable.

Activity/Exploration

Science Class: Introduce the Nuclear Fusion Model activity by showing students the nuclear fusion graphic (Figure 4.9 on p. 118). Review the components of this reaction, and ask students to reflect on how nuclear fusion differs from fission.

Nuclear Fusion Model

Student teams should begin this activity by researching fusion reactions, creating a chart, graphic organizer, or other visual means of comparing and contrasting fission and fusion reactions. For each type of reaction, students should include information about the reactants and products, the environmental impacts, safety factors, benefits, challenges, and a reflection on future use in the U.S. energy market. Then, teams should use the EDP to each create a model of a nuclear fusion reaction. Teams can use a variety of forms to create their models; for example, a model might be a computer simulation or a two-dimensional model. Students should incorporate the energy yields they calculate in mathematics class in their models.

Remind students to employ the steps of the EDP in creating their model. You may wish to have students create entries in their STEM Research Notebooks outlining how they used each step in the EDP to create their models.

Mathematics Class: Have students calculate the energy output of a fusion reaction versus a fission reaction by presenting the following problem: *The fusion reactions that are being investigated for use on Earth occur between deuterium (D) and tritium (T). Each D-T fusion reaction releases 17.6 megaelectronvolts (MeV) of energy. Each U-235 fission reaction provides 200 MeV. Convert these energy quantities into joules and express them using scientific notation. For equal masses of reactants (deuterium, tritium, and U-235), the fusion reaction produces about four times as much energy as a fission reaction. Why is this? Express the relationship between energy output in D-T fusion and U-235 fission as a mathematical formula.*

(*Note:* Students should conclude that they must consider atomic mass. Deuterium has a mass of 2.014102 amu; tritium, 3.01604927 amu; and uranium-235, 235.043924.)

ELA and Social Studies Connections: Student teams address a challenge associated with nuclear energy in the Product Development Challenge.

Product Development Challenge

Student teams of three to four students each should create a plan for a product powered by nuclear energy in response to a fictional scenario. Distribute the Product Development Challenge student handout (p. 119) and rubric (pp. 121–122), and review the scenario and challenge requirements. Student teams should each create a presentation, using either art supplies with poster or presentation board or computer software. They must provide their plan in written form as well.

Explanation

Science Class: Students need to have a conceptual understanding of nuclear force to understand fusion reactions and the conditions in which they occur. Introduce the term *Coulomb barrier* as the energy level that must be exceeded for a fusion reaction to occur. In particular, emphasize that the limitation to fusion reactions on Earth is the problem of exceeding the Coulomb barrier and generating more energy than was required as an input to initiate the reaction.

Mathematics Class: Review scientific notation and unit conversions with students.

ELA Connection: The challenge in this lesson requires students to communicate in various ways—through graphics, persuasive writing, informative writing, and oral presentations. Review these types of communication with the class, asking students to brainstorm a set of best practices for each.

Social Studies Connection: Not applicable.

Elaboration/Application of Knowledge

Science Class: Have students research various private and public enterprises that are investing in nuclear fusion and create a class database of information on the scientific and business communities' ideas about nuclear fusion. The following websites provide examples of proposed technologies and uses of nuclear fusion:

- *www.nasa.gov/directorates/spacetech/niac/slough_nuclear_propulsion.html*

- *http://generalfusion.com/what-are-the-benefits-of-fusion-energy*

- *https://lasers.llnl.gov/science/energy-for-the-future*

- *www.iter.org*

- *www.ted.com/talks/michel_laberge_how_synchronized_hammer_strikes_could_generate_ nuclear_fusion?language=en*

- *www.ted.com/talks/taylor_wilson_yup_i_built_a_nuclear_fusion_reactor*

- *www.ted.com/talks/steven_cowley_fusion_is_energy_s_future*

Mathematics Class: Have students complete the activity below.

STEM Research Notebook Prompt

Have students respond to the following prompt in their STEM Research Notebooks: *Create a mathematical model for the product you proposed in the Product Development Challenge. This model may include calculations for the amount of fuel needed to produce the desired energy output, a budget for the product's development and marketing, or both.*

ELA and Social Studies Connections: Have student teams present the business plans they created as part of the Product Development Challenge to their peers, who will act as fictional investors. Give each student in the class a fictional budget of $100 that he or she can invest in another team's innovation. Students can split this investment between multiple teams, but they cannot invest in their own team. During the presentations, have students ask questions and provide feedback. After all teams have presented, have students make their investments. Tally the investments to identify the group that received the highest investment. Discuss what was appealing about this group's product or presentation.

Evaluation/Assessment

Students may be assessed on the following performance tasks and other measures listed.

Performance Tasks

- Nuclear Fusion Model Rubric (p. 120)
- Product Development Challenge Rubric (pp. 121–122)

Other Measures

- STEM Research Notebook entries
- Collaboration Rubric (attached at the end of Lesson 2 on pp. 101–102)

INTERNET RESOURCES

Nuclear fusion

- *www.world-nuclear.org/information-library/current-and-future-generation/nuclear-fusion-power.aspx*

"What Is Fusion?" video

- *www.phdcomics.com/comics.php?f=1716*

Proposed technologies and uses of nuclear fusion

- *www.nasa.gov/directorates/spacetech/niac/slough_nuclear_propulsion.html*
- *http://generalfusion.com/what-are-the-benefits-of-fusion-energy*
- *https://lasers.llnl.gov/science/energy-for-the-future*
- *www.iter.org*
- *www.ted.com/talks/michel_laberge_how_synchronized_hammer_strikes_could_generate_nuclear_fusion?language=en*
- *www.ted.com/talks/taylor_wilson_yup_i_built_a_nuclear_fusion_reactor*
- *www.ted.com/talks/steven_cowley_fusion_is_energy_s_future*

FIGURE 4.9. NUCLEAR FUSION REACTION

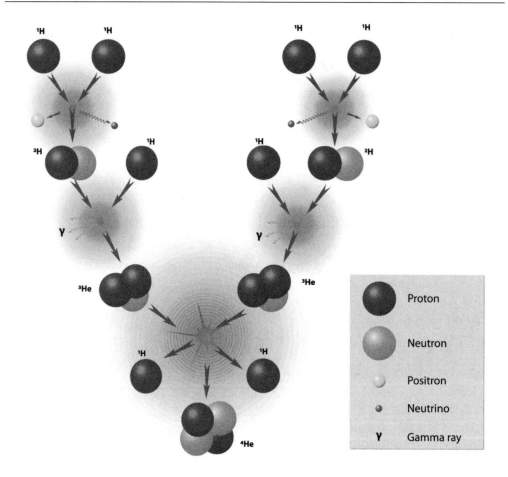

Note: A full-color version of this image is available on the book's Extras page at *www.nsta.org/roadmap-radioactivity.*

STUDENT HANDOUT

PRODUCT DEVELOPMENT CHALLENGE

Your team will work together to create a plan for a product powered by nuclear energy in response to the following scenario:

It is the year 2271, and humans have mastered the atom. Energy is readily available from radioactive decay, nuclear fission, and nuclear fusion. You are part of a group of scientists and engineers who wish to create a product that uses one of these energy sources. You want your product to be safe for humans and the environment, while being useful and marketable.

Your job therefore is to come up with an idea for an innovative product that uses one or more nuclear processes for its energy source and create a marketing plan to persuade investors to invest in your idea. The product can be big or small, something used in everyday life or for a major endeavor such as space exploration.

Your team will use the engineering design process to determine your product and assign tasks. You will provide the following business plan elements as products:

- A schematic drawing of your product that includes all parts and energy sources

- An artistic drawing of your product for marketing purposes

- A statement about the product's environmental impact and safety

- A persuasive argument about why investors should invest in your product

All team members will be responsible for researching the product and contributing their findings. Be sure to describe your product in detail and how nuclear energy is needed for it. You may use any means you wish for presenting your plan, but you must provide your plan in written form as well. Be prepared to have your plan critiqued by your classmates!

Nuclear Fusion Model Rubric					
Team Name: _____					
Type of Model	Needs Improvement (1 point)	Approaches Expectations (2 points)	Meets Expectations (3 points)	Exceeds Expectations (4 points)	*Score*
PHYSICAL MODEL	The model is missing major components and does not represent fusion reactions correctly.	The model includes the correct initial element but is lacking products or energy emissions.	The model is created from 3-D objects and includes the reactants, products, neutrons, and energy emissions associated with the reactions, but it may contain a few minor errors.	The model is created from 3-D objects and includes the reactants, products, neutrons, and energy emissions associated with the reactions.	
COMPUTER-GENERATED MODEL	The model is a single computer-generated picture of the fusion process. There may be several mistakes.	The model demonstrates the fusion process through computer-generated images or frames that are essentially pictures. There are mistakes in the model.	The model aptly demonstrates the fusion process through some form of stop-motion animation (not fluid) that shows the process of fusion, including reactants, products, and energy emissions. The initial model is accurate with few flaws.	The model aptly demonstrates the fusion process through some form of fluid animation that shows the process of fusion, including reactants, products, and energy emissions. The initial model is accurate and without flaws.	
2-D MODEL	The model is inaccurate, and there are errors in the depiction of the fusion process and energy emissions.	While the model may be created well, there are errors in the depiction of the fusion process or energy emissions.	The model is created well, with some errors in the depiction of the fusion process.	Model is hand-drawn or created and printed using computer software. It is well organized and includes the reactants, products, neutrons, and energy emissions associated with the reactions.	

TOTAL SCORE: _____

COMMENTS:

Product Development Challenge Rubric

Team Name: _____

Criteria	Needs Improvement (1 point)	Approaches Expectations (2 points)	Meets Expectations (3 points)	Exceeds Expectations (4 points)	Score
PRODUCT	The product is trite or not useful and demonstrates little understanding of nuclear energy's potential.	The product is something used presently or that was invented in the past and is only moderately improved by using nuclear energy.	The product is interesting and useful to consumers, and it adequately reflects nuclear energy's future potential.	The product imagined by the team is original and reflects a strong understanding of nuclear energy's future potential.	
SCHEMATIC DRAWING	The drawing is disorganized and difficult to understand	The drawing is somewhat organized but may not be neat, may be missing labels, or may not demonstrate students' understanding of the nuclear processes involved.	The drawing is neat and organized, with most parts labeled. It is understandable and demonstrates students' understanding of the nuclear processes involved.	The drawing is neat, organized, and fully labeled. It is easy to understand and provides details that demonstrate students' understanding of the nuclear processes involved.	
ARTISTIC DRAWING	The drawing is not neat, and it is difficult to discern how the product will appear to consumers.	The drawing may not be neat, and while it provides some details, it is unclear how the product will appear to consumers.	The drawing is neat and provides some detail to depict how the product will appear to consumers.	The drawing is neat and easy to understand, with a level of detail that provides an excellent depiction of how the product will appear to consumers.	
STATEMENT ABOUT THE PRODUCT'S ENVIRONMENTAL IMPACT AND SAFETY	The statement lacks details and clarity. Information may be inaccurate, and students' understanding of the environmental and safety concerns associated with nuclear energy is not demonstrated.	The statement has some details but may lack clarity. Some information is accurate, but it leaves questions about students' understanding of the environmental and safety concerns associated with nuclear energy.	The statement includes some details and is easy to understand. Information is accurate and demonstrates students' understanding of the environmental and safety concerns associated with nuclear energy.	The statement is detailed and easy to understand, and it provides accurate information backed up by facts. The statement demonstrates students' understanding of the environmental and safety concerns associated with nuclear energy.	

Continued

Product Development Challenge Rubric (*continued*)

Criteria	Needs Improvement (1 point)	Approaches Expectations (2 points)	Meets Expectations (3 points)	Exceeds Expectations (4 points)	Score
PERSUASIVE ARGUMENT	The argument is not convincing or is unclear and does not describe the product's benefits to consumers and investors.	The argument is convincing and sufficient in describing the product; however, the product's benefits to consumers and investors is unclear.	The argument is convincing and clear in describing the product to potential investors. It describes some benefits of the product to consumers and to investors.	The argument presented to potential investors is creative and engaging. It describes the benefits of the product to consumers and the benefits to investors.	

TOTAL SCORE: _____

COMMENTS:

Lesson Plan 4: Anatomy of a Nuclear Accident

In this lesson, students continue their consideration of nuclear fission and its practical application in power plants by examining the events leading to the 1979 Three Mile Island (TMI) nuclear accident and the response. Students analyze a multidisciplinary primary source document giving an account from the perspectives of science, mathematics, public relations, and public policy. *Note:* Although this investigation is presented as a science activity, it may also take place in mathematics, social studies, or ELA class, or in a combination of classes. Because of the interdisciplinary nature of this activity, no additional social studies and ELA activities are provided for this lesson.

ESSENTIAL QUESTIONS

- How did the TMI accident occur?

- What was the response of plant operators to the TMI accident?

- How was information conveyed throughout the accident and its aftermath?

- What are the environmental and public health implications of a nuclear accident?

- How can we limit the dangers of nuclear accidents in the future?

ESTABLISHED GOALS AND OBJECTIVES

At the conclusion of this lesson, students will be able to do the following:

- Understand the TMI nuclear accident

- Apply their understanding of nuclear fission reactions to provide a scientific explanation for the TMI accident

- Understand the responses of various stakeholder groups to the TMI nuclear accident

- Apply their understanding of the TMI nuclear accident to create a team poster providing an overview of one aspect of the accident and present it to the class

- Solve mathematical problems related to the TMI nuclear accident using dimensional analysis

TIME REQUIRED

- 4 days (approximately 45 minutes each; see Table 3.9, p. 40)

MATERIALS

Required Materials for Lesson 4

- STEM Research Notebooks

- Computers with internet access for student research and viewing videos (for each team)

- Handouts (attached at the end of this lesson)

Additional Materials for Anatomy of an Accident (per team unless otherwise noted)

- Printed or electronic copy of "Account of the Accident," pages 101–161 of the *Report of the President's Commission on the Three Mile Island Accident* (1 per student; *https://catalog.hathitrust.org/Record/007418765*)

- Anatomy of an Accident Study Guide (pp. 138–139; 1 per student)

- Poster board

- Set of markers

- Glue

- Scissors

- Colored paper (6 sheets)

SAFETY NOTES

1. Use caution when working with sharps (scissors) to avoid cutting or puncturing skin.

2. Wash hands with soap and water after the activity is completed.

CONTENT STANDARDS AND KEY VOCABULARY

Table 4.10 lists the content standards from the *NGSS*, *CCSS*, and the Framework for 21st Century Learning that this lesson addresses, and Table 4.11 (p. 130) presents the key vocabulary. Vocabulary terms are provided for both teacher and student use. Teachers may choose to introduce some or all of the terms to students.

Table 4.10. Content Standards Addressed in STEM Road Map Module
Lesson 4

NEXT GENERATION SCIENCE STANDARDS

PERFORMANCE EXPECTATIONS

- HS-PS1-1. Use the periodic table as a model to predict the relative properties of elements based on the patterns of electrons in the outermost energy level of atoms.

- HS-PS1-7. Use mathematical representations to support the claim that atoms, and therefore mass, are conserved during a chemical reaction.

- HS-PS1-8. Develop models to illustrate the changes in the composition of the nucleus of the atom and the energy released during the processes of fission, fusion, and radioactive decay.

- HS-ETS-2. Design a solution to a complex real-world problem by breaking it down into smaller, more manageable problems that can be solved through engineering.

SCIENCE AND ENGINEERING PRACTICES

Using Mathematics and Computational Thinking

- Use mathematical representations of phenomena to support claims.

- Create a computational model or simulation of a phenomenon, designed device, process, or system.

Developing and Using Models

- Use a model to predict the relationships between systems or between components of a system.

- Develop a model based on evidence to illustrate the relationships between systems or between components of a system.

Obtaining and Communicating Information

- Evaluate the validity and reliability of multiple claims that appear in scientific and technical texts or media reports, verifying the data when possible.

- Communicate technical information or ideas in multiple formats.

Constructing Explanations and Designing Solutions

- Design, evaluate, and/or refine a solution to a complex real-world problem, based on scientific knowledge, student-generated sources of evidence, prioritized criteria, and trade-off considerations.

- Construct and revise an explanation based on valid and reliable evidence obtained from a variety of sources (including students' own investigations, models, theories, simulations, peer review) and the assumption that theories and laws that describe the natural world operate today as they did in the past and will continue to do so in the future.

Continued

Table 4.10. (*continued*)

DISCIPLINARY CORE IDEAS

PS1.A: Structure and Properties of Matter

- Each atom has a charged substructure consisting of a nucleus, which is made of protons and neutrons, surrounded by electrons.

- The periodic table orders elements horizontally by the number of protons in the atom's nucleus and places those with similar chemical properties in columns. The repeating patterns of this table reflect patterns of outer electron states.

- A stable molecule has less energy than the same set of atoms separated; one must provide at least this energy in order to take the molecule apart.

PS1.B: Chemical Reactions

- The fact that atoms are conserved, together with knowledge of the chemical properties of the elements involved, can be used to describe and predict chemical reactions.

PS1.C: Nuclear Processes

- Nuclear processes, including fusion, fission, and radioactive decays of unstable nuclei, involve release or absorption of energy. The total number of neutrons plus protons does not change in any nuclear process.

- Spontaneous radioactive decays follow a characteristic exponential decay law. Nuclear lifetimes allow radiometric dating to be used to determine the ages of rocks and other materials.

PS2.B: Types of Interactions

- Attraction and repulsion between electric charges at the atomic scale explain the structure, properties, and transformations of matter, as well as the contact forces between material objects.

PS3.B: Conservation of Energy and Energy Transfer

- Conservation of energy means that the total change of energy in any system is always equal to the total energy transferred into or out of the system.

- Energy cannot be created or destroyed, but it can be transported from one place to another and transferred between systems.

- Mathematical expressions, which quantify how the stored energy in a system depends on its configuration (e.g., relative positions of charged particles, compression of a spring) and how kinetic energy depends on mass and speed, allow the concept of conservation of energy to be used to predict and describe system behavior.

- The availability of energy limits what can occur in any system.

Continued

Table 4.10. (*continued*)

PS3.D: Energy in Chemical Processes
- Although energy cannot be destroyed, it can be converted to less useful forms—for example, to thermal energy in the surrounding environment.

ETS1.A: Defining and Delimiting an Engineering Problem
- Criteria and constraints also include satisfying any requirements set by society, such as taking issues of risk mitigation into account, and they should be quantified to the extent possible and stated in such a way that one can tell if a given design meets them.

ETS1.C: Optimizing the Design Solution
- Criteria may need to be broken down into simpler ones that can be approached systematically, and decisions about the priority of certain criteria over others (trade-offs) may be needed.

CROSSCUTTING CONCEPTS

Patterns
- Different patterns may be observed at each of the scales at which a system is studied and can provide evidence for causality in explanations of phenomena.

Energy and Matter
- Changes of energy and matter in a system can be described in terms of energy and matter flows into, out of, and within that system.
- In nuclear processes, atoms are not conserved, but the total number of protons plus neutrons is conserved.
- The total amount of energy and matter in closed systems is conserved.

Stability and Change
- Much of science deals with constructing explanations of how things change and how they remain stable.

Cause and Effect
- Cause and effect relationships can be suggested and predicted for complex natural and human designed systems by examining what is known about smaller scale mechanisms within the system.

Scale, Proportion, and Quantity
- The significance of a phenomenon is dependent on the scale, proportion, and quantity at which it occurs.

Systems and System Models
- Models can be used to predict the behavior of a system, but these predictions have limited precision and reliability due to the assumptions and approximations inherent in models.

Continued

Table 4.10. (*continued*)

CROSSCUTTING CONCEPTS (*continued*)

Structure and Function

- Investigating or designing new systems or structures requires a detailed examination of the properties of different materials, the structures of different components, and connections of components to reveal its function and/or solve a problem.

COMMON CORE STATE STANDARDS FOR MATHEMATICS

MATHEMATICAL PRACTICES

- MP1. Make sense of problems and persevere in solving them.
- MP3. Construct viable arguments and critique the reasoning of others.
- MP4. Model with mathematics.
- MP7. Look for and make use of structure.

MATHEMATICAL CONTENT

- HSF.IF.B.6. Calculate and interpret the average rate of change of a function (presented symbolically or as a table) over a specified interval. Estimate the rate of change from a graph.
- HSF.LE.B.5. Interpret the parameters in a linear or exponential function in terms of a context.
- HSF.BF.A.1. Write a function that describes a relationship between two quantities.
- HSF.BF.A.2. Write arithmetic and geometric sequences both recursively and with an explicit formula, use them to model situations, and translate between the two forms.
- HSF.BF.B.5. Understand the inverse relationship between exponents and logarithms and use this relationship to solve problems involving logarithms and exponents.

COMMON CORE STATE STANDARDS FOR ENGLISH LANGUAGE ARTS

READING STANDARDS

- RI.11-12.1. Cite strong and thorough textual evidence to support analysis of what the text says explicitly as well as inferences drawn from the text, including determining where the text leaves matters uncertain.
- RI.11-12.2. Determine two or more central ideas of a text and analyze their development over the course of the text, including how they interact and build on one another to provide a complex analysis; provide an objective summary of the text.
- RI.11-12.3. Analyze a complex set of ideas or sequence of events and explain how specific individuals, ideas, or events interact and develop over the course of the text.
- RI.11-12.4. Determine the meaning of words and phrases as they are used in a text, including figurative, connotative, and technical meanings; analyze how an author uses and refines the meaning of a key term or terms over the course of a text.

Continued

Table 4.10. (*continued*)

READING STANDARDS (*continued*)

- RI.11-12.5. Analyze and evaluate the effectiveness of the structure an author uses in his or her exposition or argument, including whether the structure makes points clear, convincing, and engaging.

- RI.11-12.7. Integrate and evaluate multiple sources of information presented in different media or formats (e.g., visually, quantitatively) as well as in words in order to address a question or solve a problem.

- RI.11-12.8. Delineate and evaluate the reasoning in seminal U.S. texts, including the application of constitutional principles and use of legal reasoning and the premises, purposes, and arguments in works of public advocacy.

WRITING STANDARDS

- W.11-12.1. Write arguments to support claims in an analysis of substantive topics or texts, using valid reasoning and relevant and sufficient evidence.

- W.11-12.2. Write informative/explanatory texts to examine and convey complex ideas, concepts, and information clearly and accurately through the effective selection, organization, and analysis of content.

SPEAKING AND LISTENING STANDARDS

- SL.11-12.1. Initiate and participate effectively in a range of collaborative discussions (one-on-one, in groups, and teacher-led) with diverse partners on grades 11–12 topics, texts, and issues, building on others' ideas and expressing their own clearly and persuasively.

- SL.11-12.2. Integrate multiple sources of information presented in diverse formats and media (e.g., visually, quantitatively, orally) in order to make informed decisions and solve problems, evaluating the credibility and accuracy of each source and noting any discrepancies among the data.

- SL.11-12.3. Evaluate a speaker's point of view, reasoning, and use of evidence and rhetoric, assessing the stance, premises, links among ideas, word choice, points of emphasis, and tone used.

- SL.11-12.4. Present information, findings, and supporting evidence, conveying a clear and distinct perspective, such that listeners can follow the line of reasoning, alternative or opposing perspectives are addressed, and the organization, development, substance, and style are appropriate to purpose, audience, and a range of formal and informal tasks.

- SL.11-12.5. Make strategic use of digital media (e.g., textual, graphical, audio, visual, and interactive elements) in presentations to enhance understanding of findings, reasoning, and evidence and to add interest.

- SL.11-12.6. Adapt speech to a variety of contexts and tasks, demonstrating a command of formal English when indicated or appropriate.

Continued

Table 4.10. (*continued*)

FRAMEWORK FOR 21ST CENTURY LEARNING
• Interdisciplinary Themes: Global Awareness; Financial, Economic, Business and Entrepreneurial Literacy; Civic Literacy; Environmental Literacy
• Learning and Innovation Skills: Creativity and Innovation, Critical Thinking and Problem Solving, Communication and Collaboration
• Information, Media, and Technology Skills: Information Literacy, Media Literacy, ICT Literacy
• Life and Career Skills: Flexibility and Adaptability, Initiative and Self-Direction, Social and Cross-Cultural Skills, Productivity and Accountability, Leadership and Responsibility

Table 4.11. Key Vocabulary for Lesson 4

Key Vocabulary	Definition
Roentgen equivalent man (rem)	a unit used to measure radiation exposure that accounts for the amount of ionizing radiation and the type of radiation (alpha, beta, gamma, or neutron); biological doses are commonly measured in millirems (mrem)
Geiger counter	a tool commonly used for measuring ionizing radiation; can measure either counts (the number of ionizing radiation hits) or radiation dose, but cannot distinguish among types of radiation so cannot provide information about rem
sievert (sv)	the International System (SI) unit for measuring exposure to ionizing radiation equal to 1 joule/kilogram (1 sv = 100 rem)

TEACHER BACKGROUND INFORMATION
Three Mile Island Nuclear Accident

For this lesson, you should have a basic understanding of the 1979 Three Mile Island (TMI) nuclear accident, which was the worst nuclear accident in U.S. history. Human error and mechanical problems combined to create conditions that were unexpected and for which operators were not prepared. Although the environmental and human impacts of the accident were ultimately determined to be minimal, the Nuclear Regulatory Commission (NRC) responded with extensive regulations that have had lasting impact on the nuclear energy industry.

Students' investigation of the chain of events in this accident will provide important information for them as they undertake their final challenge in the next lesson. Students will use a primary source, *Report of the President's Commission on the Accident at Three Mile Island* (Kemeny et al. 1979), a full copy of which can be found at the following website:

https://catalog.hathitrust.org/Record/007418765. In particular, students will focus on the section "Account of the Accident," a chronological, hour-by-hour, diary-type description that provides insight into the experiences of the people involved and their attempts to cope with the many implications of the accident. You may also wish to review the NRC's TMI background report and associated video at *www.nrc.gov/reading-rm/doc-collections/ fact-sheets/3mile-isle.html.* The Smithsonian National Museum of American History has information about the event and drawings of the reactor core at various points throughout the accident at *www.americanhistory.si.edu/tmi/tmi03.htm.*

COMMON MISCONCEPTIONS

Students will have various types of prior knowledge about the concepts introduced in this lesson. Table 4.12 outlines some common misconceptions students may have concerning these concepts. Because of the breadth of students' experiences, it is not possible to anticipate every misconception that students may bring as they approach this lesson. Incorrect or inaccurate prior understanding of concepts can influence student learning in the future, however, so it is important to be alert to misconceptions such as those presented in the table.

Table 4.12. Common Misconceptions About the Concepts in Lesson 4

Topic	Student Misconception	Explanation
Nuclear science and technology	Having a long half-life means that a radioactive substance is especially dangerous to humans.	The length of an element's half-life indicates only how fast the material decays and not its potential to harm humans; the critical factor to consider is the level of radioactivity in an element. For example, carbon-14 has a long half-life (5,730 years) but is present in all living things and is not harmful; conversely, nitrogen-16 has a very short half-life (7.1 seconds) but emits high-energy radiation that is harmful to humans.
	A nuclear power plant can explode like an atomic bomb.	Reactor fuel does not contain enough uranium to cause an explosion; in addition, plants are built with many safety controls and features that make the nuclear reaction self-limiting. It is not possible for people to modify nuclear power plants to turn them into atomic weapons.
	The "smoke" coming out of the "smokestacks" at nuclear power plants contains harmful emissions.	The "smokestacks" are actually cooling towers, and the white "smoke" seen coming out of the towers at nuclear power plants is steam from water that has not come into contact with any radioactive material.

PREPARATION FOR LESSON 4

Review the Teacher Background Information section (p. 130), assemble the materials for the lesson, make copies of the student handout, and preview the videos recommended in the Learning Components section. Make arrangements for a presentation period or evening event if you wish to have students present their Anatomy of an Accident posters to a larger audience (e.g., parents, school administrators, other classes).

LEARNING COMPONENTS
Introductory Activity/Engagement

Connection to the Challenge: Begin each day of this lesson by directing students' attention to the driving question for the module and challenge: How does the use of nuclear energy to meet our energy demands affect society? Hold a brief discussion of how students' learning in the previous days' lessons contributed to their ability to create their response to the Gammatown crisis. You may wish to hold a class discussion, creating a class list of key ideas on chart paper, or have students create a STEM Research Notebook entry with this information.

Science Class: Show students the ABC News report "Three Mile Island, Nuclear Power Plant Accident: March 28, 1979" at *www.youtube.com/watch?v=eGl7VymjSho*. Hold a class discussion about the report, asking students to respond to the following questions:

- What information does this report provide?

- What important information is missing from this report?

- Was there cause for public concern?

- Why do you think the alarm was not sounded until three hours after the accident?

- Would you be worried if you lived within 10 miles of the reactor?

Next, watch the March 30, 1979, ABC News report "Three Mile Island, Safety Fears in 1979" at *www.youtube.com/watch?v=2VRdkTvv878*. Ask students to reflect on the differences between the two reports in the information presented and the tone. Hold a class discussion, asking the following questions:

- What new information did you get from this report?

- What questions would you have if you lived near the plant?

- Do you agree with the governor's statement that "there is no imminent danger"?

- The reporter said, "The first casualty of this accident may have been trust." What does he mean by this? Trust in whom or what? Why is trust important?

- A man in the report expressed concern about effects up to 25 years after. Was this concern valid? Why or why not?

(*Note:* Radioactive exposure can result in delayed effects due to the potential for genetic damage. If a radioactive particle injures a DNA molecule, that injury is propagated in all future generations of that cell. For example, if the cell becomes a cancer cell, all future cells propagated from that cell will be cancerous; if the cell is an egg or sperm cell, all children born from that egg or sperm will carry the altered DNA.)

Mathematics Class: Introduce the units used to measure exposure to ionizing radiation, sievert (sv) and Roentgen equivalent man (rem). Tell students that a lethal dose of radiation is considered to be around 400 rem with exposure occurring over a short period of time. Show students a summary of the effects of radiation on human health (an example is the chart found at *www.atomicarchive.com/Effects/radeffectstable.shtml*).

Ask students if they believe any of the following would result in exposure to radiation and, if so, how much radiation in rems:

- 3 days of living in Atlanta

- 2 days of living in Denver

- 1 year of watching television (on average)

- 1 year of wearing a watch with a luminous dial

- 1 coast-to-coast airline flight

- 1 year living next door to a normally operating nuclear power plant

In fact, all these activities result in exposure to 1 mrem, or 0.01 rem. Each 1 mrem exposure is theorized to reduce life expectancy by about 1.2 minutes. Based on this, ask students to calculate the number of rems that would reduce their life expectancy by one year.

Have students use the NRC's Personal Annual Radiation Dose Calculator at *www. nrc.gov/about-nrc/radiation/around-us/doses-daily-lives.html#3* to find their personal yearly dose of radiation. (You may wish to hand out copies of the printable worksheet provided on the web page as an alternative to the online calculator.)

ELA Connection: Not applicable.

Social Studies Connection: Not applicable.

Activity/Exploration

Science Class: In the Anatomy of an Accident activity, students explore the events surrounding the TMI nuclear accident using the "Account of the Accident" from *Report of the President's Commission on the Accident at Three Mile Island* (Kemeny et al. 1979).

Anatomy of an Accident

This activity should be undertaken using a jigsaw approach in which small groups take responsibility for recording and reporting on various topics. The goal of this activity is to create a comprehensive visual guide to the TMI accident that includes the following:

- A flow chart of events in the power plant during the accident

- A flow chart of events in the community and media during the accident

- Log of human error during the accident that can be correlated with the flow charts

- Log of when radioactivity escaped the reactor, including the amounts, times, and places it was detected, and a diagram of the reactor, containment building, and auxiliary building indicating where the mechanical failures took place

- A "cast of characters" listing the key players involved, their jobs, and their contributions during the crisis

You may wish to have students read the account aloud in class while student groups act as "scribes," recording information for their topics by taking notes or using graphic organizers. Alternatively, you might assign portions of the reading for homework to allow for additional group work time in class. Distribute copies of the Anatomy of an Accident Study Guide handout (pp. 138–139), which will help students focus their reading and gather important information about the event.

After the reading is complete, each team should create an informative, easy-to-read, visually appealing poster displaying the information for its assigned topic. Each group should then present its poster to the entire class, highlighting important events and information.

Mathematics Class: Introduce students to the dose-response model currently accepted by the NRC, the linear no-threshold (LNT) dose-response relationship model, which describes the relationship between exposure to ionizing radiation and the occurrence of cancer and suggests that any increase in dose represents an increase in risk (see *www. nrc.gov/about-nrc/radiation/health-effects/rad-exposure-cancer.html* for more information). This model has been criticized, particularly among businesses that must protect their employees against low levels of radiation. Some researchers have suggested alternative models. These include a hermetic model, contending that low doses of radiation, below a particular threshold, are benign or even promote health by stimulating the immune system. Ask students to reflect on why some organizations might oppose the LNT model.

Have student teams work together to create graphs that depict the relationships between radiation and health risks presented by the two models. Next, have teams

formulate equations for their graphs, and then have them share their graphs and equations with the class. Discuss the differences and similarities among teams' responses.

ELA Connection: Not applicable.

Social Studies Connection: Not applicable.

Explanation

Science Class: Students may need some background about the context in which the TMI nuclear accident took place. Have students locate the TMI site (near Harrisburg, Pennsylvania) on a map. As a class, discuss the 1970s energy crisis and the influence that had on the climate for nuclear energy. You may wish to have students explore the History Channel's "Energy Crisis (1970s)" at *www.history.com/topics/energy-crisis*.

Mathematics Class: Review linear and exponential functions and the differences between the two using graphic representations and equations. Provide several scenarios of human radiation exposure, and have student teams apply the equations they created for the LNT and hermetic models to identify the suggested health risks. Have students compare the teams' findings and discuss the public health and safety implications of each model. Discuss how health and safety regulations might change if the NRC adopted a hermetic model rather than the LNT model it currently uses to guide safety regulations.

ELA Connection: Not applicable.

Social Studies Connection: Not applicable.

Elaboration/Application of Knowledge

Science Class: Have students complete the activity below.

STEM Research Notebook Prompt

Have students respond to the following prompt in their STEM Research Notebooks, and then hold a class discussion: *When it became safe to open the reactor at Three Mile Island years after the accident, it became apparent that the reactor had become nearly 90% uncovered and was much closer to a nuclear meltdown than was originally thought. If this information had become available sooner, how might the response to the accident have been different, both locally and at a national level?*

Hold a class discussion in which students share and discuss their responses to this STEM Research Notebook prompt. You may wish to have the teams present their TMI Anatomy of an Accident posters to a broader audience (e.g., parents, school administrators, other classes).

Mathematics Class: Have students calculate the radiation of varying quantities of iodine-131 (in grams), one of the radioactive isotopes that was released during the TMI accident. I-131 has a half-life of 8 days. Their challenge is to determine how much radioactivity was released using the equation $a = n_0 / (1.44 \times t_{1/2})$, where a = dissociations per second (or becquerels [Bq]), n_0 = the number of atomic nuclei, and $t_{1/2}$ = half-life in seconds. Students will need to convert grams to atoms by using dimensional analysis.

Next, tell students that about 560 gigabecquerels (GBq) of I-131 were released during the TMI accident. Have them calculate how many grams of I-131 this represents by using the equation above (1 GBq = 1,000,000,000 or 10^9 becquerels).

ELA Connection: Not applicable.

Social Studies Connection: Not applicable.

Evaluation/Assessment

Students may be assessed on the following performance tasks and other measures listed.

Performance Tasks

- Anatomy of an Accident team posters and presentations

- Anatomy of an Accident Study Guide (pp. 138–139)

Other Measures

- STEM Research Notebook entries

- Collaboration Rubric (attached at the end of Lesson 2 on pp. 101–102)

INTERNET RESOURCES

Report of the President's Commission on the Accident at Three Mile Island (Kemeny et al. 1979)
- *https://catalog.hathitrust.org/Record/007418765*

NRC's TMI background report
- *www.nrc.gov/reading-rm/doc-collections/fact-sheets/3mile-isle.html*

TMI historical overview
- *www.americanhistory.si.edu/tmi/tmi03.htm*

News reports of the TMI accident
- *www.youtube.com/watch?v=eGI7VymjSho*

- *www.youtube.com/watch?v=2VRdkTvv878*

Effects of radiation on human health
- *www.atomicarchive.com/Effects/radeffectstable.shtml*

Personal Annual Radiation Dose Calculator
- *www.nrc.gov/about-nrc/radiation/around-us/doses-daily-lives.html#3*

LNT dose-response model
- *www.nrc.gov/about-nrc/radiation/health-effects/rad-exposure-cancer.html*

1970s energy crisis
- *www.history.com/topics/energy-crisis*

REFERENCE

Kemeny, J. G., B. Babbitt, P. E. Haggerty, C. Lewis, P. A. Marks, C. B. Marrett, L. McBride, H. C. McPherson, R. W. Peterson, T. H. Pigford, T. B. Taylor, and A. D. Trunk. 1979. *Report of the President's Commission on the accident at Three Mile Island.* Washington, DC: U.S. Government Printing Office. *https://catalog.hathitrust.org/Record/007418765.*

STUDENT HANDOUT, PAGE 1

ANATOMY OF AN ACCIDENT STUDY GUIDE

Read "Account of the Accident" from *Report of the President's Commission on the Accident at Three Mile Island* (*https://catalog.hathitrust.org/Record/007418765*) and answer these questions on a separate sheet of paper.

1. List the details of the Three Mile Island (TMI) power plant (e.g., where it is, how many reactors it has, how much energy it produces).
2. What are the elements of the TMI system during normal operations?
3. What are the three safety barriers in nuclear power plants?
4. How do the control rods work?
5. How tall was the core of the reactor?
6. What happens if the reactor's core becomes uncovered?
7. What is the melting point of uranium fuel?
8. What causes the white vapor that comes from the cooling towers at nuclear power plants?
9. Why didn't the Emergency Core Cooling System work?
10. What are the functions of steam in a nuclear reactor?
11. Why didn't the reactor stop generating heat after the control rods were dropped into it?
12. Who were the four people who dealt with the early stages of the accident?
13. Who produced the reactor?
14. What is a polisher?
15. What were the indications throughout the event that the reactor's water was boiling and becoming steam?
16. Why was hydrogen produced and how did it get into the containment building?
17. When did the operators realize that they faced a serious emergency at the reactor?
18. What were the radiation levels in the containment building at 6:30 a.m.? At 7:20 a.m.?
19. Who took the role of emergency director?
20. What was the definition of a "general emergency"?
21. Where was radiation detected at 7:48 a.m.? How much?
22. What is isolation?
23. What were the 1975 criteria for isolation? Were these put into effect at TMI? Why or why not?

ANATOMY OF AN ACCIDENT STUDY GUIDE

24. How did the mayor of Harrisburg learn about the accident?

25. What type of radiation was detected in Goldsboro, Pennsylvania, the day of the accident?

26. Who provided information to the public about the accident?

27. What information did Lieutenant Governor Scranton provide in his press conference on the morning of March 28?

28. What information did Lieutenant Governor Scranton provide in his press conference on the afternoon of March 28? Was the information he provided accurate? If not, how was it inaccurate?

29. At what time did the hydrogen explosion occur? Where did the explosion occur?

30. What were the radiation readings in Middletown the afternoon of March 28?

31. When did the U.S. Food and Drug Administration begin monitoring food, milk, and water in the area surrounding TMI?

32. When was water from the plant released into the Susquehanna River? What was the source of the water? What was the radioactivity level?

33. What was the cause of the 8 a.m. radiation releases over the containment building?

34. When was an evacuation ordered? Who was evacuated?

35. What event reassured the Harrisburg mayor on Friday, March 30?

36. When did the nuclear industry become involved? How was it involved?

37. What were officials from the Department of Health, Education, and Welfare searching for?

38. What is radiolysis?

39. Why didn't the local emergency preparedness office have current information to provide to citizens on April 1?

40. Why wasn't NRC official Victor Stello concerned about oxygen building up inside the reactor?

41. How did the hydrogen bubble diminish?

42. What was the intended tone of the April 2 morning press conference?

43. Why did Gordon MacLeod, Pennsylvania's secretary of health, oppose distributing potassium iodide to the public?

44. When did the governor withdraw the evacuation advisory?

Lesson Plan 5: The Gammatown Crisis Challenge

In this lesson, students synthesize their learning from the previous lessons to address the module's culminating challenge, the Gammatown Crisis Challenge. Student teams are each challenged to assume the role of one of several stakeholder groups to create a response to a fictional nuclear accident in Gammatown. Students prepare a presentation, printed material, and a physical or mathematical model relevant to the target audience of their team's stakeholder group. Then, they present these materials during a town hall meeting.

ESSENTIAL QUESTIONS

- How can we inform citizens and government officials about the implications of a nuclear accident?

- How can we prevent future nuclear accidents?

ESTABLISHED GOALS AND OBJECTIVES

At the conclusion of this lesson, students will be able to do the following:

- Apply their understanding of nuclear energy to understand stakeholder-specific responses to a nuclear accident

- Demonstrate understanding of nuclear science in the presentation

- Demonstrate understanding of the mathematics concepts introduced in the module

- Use mathematical modeling to convey information about nuclear reactions

- Use persuasive language to present an argument to a target audience

- Apply the EDP to solve a complex problem

- Collaborate with peers to solve a problem

- Create a prototype related to a science or technology aspect of nuclear energy

TIME REQUIRED

- 6 days (approximately 45 minutes each; see Tables 3.9 and 3.10, pp. 40–41)

MATERIALS

Required Materials for Lesson 5

- STEM Research Notebooks

- Computers with internet access (for each team)

- Presentation software (for teams who choose to create presentations using software)

- Video recording equipment (for teams who choose to create presentations using video)

- Supplies for printed materials to be determined by each team (e.g., poster board, presentation board, white paper, markers, word processing or graphic design program)

- Supplies to create prototype (to be provided by students)

- Indirectly vented chemical splash safety goggles

- Handouts (attached at the end of this lesson)

SAFETY NOTES

1. All students must wear safety goggles during all phases of this inquiry activity.

2. Use caution when working with sharps (e.g., scissors) to avoid cutting or puncturing skin.

3. Wash hands with soap and water after the activity is completed.

CONTENT STANDARDS AND KEY VOCABULARY

Table 4.13 (p. 142) lists the content standards from the *NGSS, CCSS,* and the Framework for 21st Century Learning that this lesson addresses, and Table 4.14 (p. 146) presents the key vocabulary. Vocabulary terms are provided for both teacher and student use. Teachers may choose to introduce some or all of the terms to students.

Table 4.13. Content Standards Addressed in STEM Road Map Module Lesson 5

NEXT GENERATION SCIENCE STANDARDS

PERFORMANCE EXPECTATIONS

- HS-PS1-1. Use the periodic table as a model to predict the relative properties of elements based on the patterns of electrons in the outermost energy level of atoms.

- HS-PS1-7. Use mathematical representations to support the claim that atoms, and therefore mass, are conserved during a chemical reaction.

- HS-PS1-8. Develop models to illustrate the changes in the composition of the nucleus of the atom and the energy released during the processes of fission, fusion, and radioactive decay.

- HS-ETS-2. Design a solution to a complex real-world problem by breaking it down into smaller, more manageable problems that can be solved through engineering.

SCIENCE AND ENGINEERING PRACTICES

Using Mathematics and Computational Thinking

- Use mathematical representations of phenomena to support claims.

- Create a computational model or simulation of a phenomenon, designed device, process, or system.

Developing and Using Models

- Use a model to predict the relationships between systems or between components of a system.

- Develop a model based on evidence to illustrate the relationships between systems or between components of a system.

Obtaining and Communicating Information

- Evaluate the validity and reliability of multiple claims that appear in scientific and technical texts or media reports, verifying the data when possible.

- Communicate technical information or ideas in multiple formats.

Constructing Explanations and Designing Solutions

- Design, evaluate, and/or refine a solution to a complex real-world problem, based on scientific knowledge, student-generated sources of evidence, prioritized criteria, and trade-off considerations.

- Construct and revise an explanation based on valid and reliable evidence obtained from a variety of sources (including students' own investigations, models, theories, simulations, peer review) and the assumption that theories and laws that describe the natural world operate today as they did in the past and will continue to do so in the future.

Continued

Table 4.13. (*continued*)

DISCIPLINARY CORE IDEAS

PS1.B: Chemical Reactions

- The fact that atoms are conserved, together with knowledge of the chemical properties of the elements involved, can be used to describe and predict chemical reactions.

PS1.C: Nuclear Processes

- Nuclear processes, including fusion, fission, and radioactive decays of unstable nuclei, involve release or absorption of energy. The total number of neutrons plus protons does not change in any nuclear process.

- Spontaneous radioactive decays follow a characteristic exponential decay law. Nuclear lifetimes allow radiometric dating to be used to determine the ages of rocks and other materials.

PS3.B: Conservation of Energy and Energy Transfer

- Conservation of energy means that the total change of energy in any system is always equal to the total energy transferred into or out of the system.

- Energy cannot be created or destroyed, but it can be transported from one place to another and transferred between systems.

- Mathematical expressions, which quantify how the stored energy in a system depends on its configuration (e.g., relative positions of charged particles, compression of a spring) and how kinetic energy depends on mass and speed, allow the concept of conservation of energy to be used to predict and describe system behavior.

- The availability of energy limits what can occur in any system.

PS3.D: Energy in Chemical Processes

- Although energy cannot be destroyed, it can be converted to less useful forms—for example, to thermal energy in the surrounding environment.

ETS1.A: Defining and Delimiting an Engineering Problem

- Criteria and constraints also include satisfying any requirements set by society, such as taking issues of risk mitigation into account, and they should be quantified to the extent possible and stated in such a way that one can tell if a given design meets them.

ETS1.C. Optimizing the Design Solution

- Criteria may need to be broken down into simpler ones that can be approached systematically, and decisions about the priority of certain criteria over others (trade-offs) may be needed.

Continued

Table 4.13. (*continued*)

CROSSCUTTING CONCEPTS

Patterns

- Different patterns may be observed at each of the scales at which a system is studied and can provide evidence for causality in explanations of phenomena.

Energy and Matter

- Changes of energy and matter in a system can be described in terms of energy and matter flows into, out of, and within that system.

- In nuclear processes, atoms are not conserved, but the total number of protons plus neutrons is conserved.

- The total amount of energy and matter in closed systems is conserved.

Stability and Change

- Much of science deals with constructing explanations of how things change and how they remain stable.

Cause and Effect

- Cause and effect relationships can be suggested and predicted for complex natural and human designed systems by examining what is known about smaller scale mechanisms within the system.

Scale, Proportion, and Quantity

- The significance of a phenomenon is dependent on the scale, proportion, and quantity at which it occurs.

Systems and System Models

- Models can be used to predict the behavior of a system, but these predictions have limited precision and reliability due to the assumptions and approximations inherent in models.

Structure and Function

- Investigating or designing new systems or structures requires a detailed examination of the properties of different materials, the structures of different components, and connections of components to reveal its function and/or solve a problem.

COMMON CORE STATE STANDARDS FOR MATHEMATICS

MATHEMATICAL PRACTICES

- MP1. Make sense of problems and persevere in solving them.

- MP3. Construct viable arguments and critique the reasoning of others.

- MP4. Model with mathematics.

- MP7. Look for and make use of structure.

Continued

Table 4.13. (*continued*)

MATHEMATICAL CONTENT

- HSF.LE.A.1. Distinguish between situations that can be modeled with linear functions and with exponential functions.

- HSF.LE.A.2. Construct linear and exponential functions, including arithmetic and geometric sequences, given a graph, a description of a relationship, or two input-output pairs (include reading these from a table).

- HSF.LE.B.5. Interpret the parameters in a linear or exponential function in terms of a context.

- HSF.BF.A.1. Write a function that describes a relationship between two quantities.

COMMON CORE STATE STANDARDS FOR ENGLISH LANGUAGE ARTS

READING STANDARDS

- RI.11-12.1. Cite strong and thorough textual evidence to support analysis of what the text says explicitly as well as inferences drawn from the text, including determining where the text leaves matters uncertain.

- RI.11-12.2. Determine two or more central ideas of a text and analyze their development over the course of the text, including how they interact and build on one another to provide a complex analysis; provide an objective summary of the text.

- RI.11-12.3. Analyze a complex set of ideas or sequence of events and explain how specific individuals, ideas, or events interact and develop over the course of the text.

- RI.11-12.4. Determine the meaning of words and phrases as they are used in a text, including figurative, connotative, and technical meanings; analyze how an author uses and refines the meaning of a key term or terms over the course of a text.

- RI.11-12.5. Analyze and evaluate the effectiveness of the structure an author uses in his or her exposition or argument, including whether the structure makes points clear, convincing, and engaging.

- RI.11-12.7. Integrate and evaluate multiple sources of information presented in different media or formats (e.g., visually, quantitatively) as well as in words in order to address a question or solve a problem.

- RI.11-12.8. Delineate and evaluate the reasoning in seminal U.S. texts, including the application of constitutional principles and use of legal reasoning and the premises, purposes, and arguments in works of public advocacy.

WRITING STANDARDS

- W.11-12.1. Write arguments to support claims in an analysis of substantive topics or texts, using valid reasoning and relevant and sufficient evidence.

- W.11-12.2. Write informative/explanatory texts to examine and convey complex ideas, concepts, and information clearly and accurately through the effective selection, organization, and analysis of content.

Continued

Table 4.13. (*continued*)

> **SPEAKING AND LISTENING STANDARDS**
> - SL.11-12.1. Initiate and participate effectively in a range of collaborative discussions (one-on-one, in groups, and teacher-led) with diverse partners on grades 11–12 topics, texts, and issues, building on others' ideas and expressing their own clearly and persuasively.
>
> - SL.11-12.2. Integrate multiple sources of information presented in diverse formats and media (e.g., visually, quantitatively, orally) in order to make informed decisions and solve problems, evaluating the credibility and accuracy of each source and noting any discrepancies among the data.
>
> - SL.11-12.3. Evaluate a speaker's point of view, reasoning, and use of evidence and rhetoric, assessing the stance, premises, links among ideas, word choice, points of emphasis, and tone used.
>
> - SL.11-12.4. Present information, findings, and supporting evidence, conveying a clear and distinct perspective, such that listeners can follow the line of reasoning, alternative or opposing perspectives are addressed, and the organization, development, substance, and style are appropriate to purpose, audience, and a range of formal and informal tasks.
>
> - SL.11-12.5. Make strategic use of digital media (e.g., textual, graphical, audio, visual, and interactive elements) in presentations to enhance understanding of findings, reasoning, and evidence and to add interest.
>
> - SL.11-12.6. Adapt speech to a variety of contexts and tasks, demonstrating a command of formal English when indicated or appropriate.
>
> **FRAMEWORK FOR 21ST CENTURY LEARNING**
> - Interdisciplinary Themes: Global Awareness; Financial, Economic, Business and Entrepreneurial Literacy; Civic Literacy; Environmental Literacy
>
> - Learning and Innovation Skills: Creativity and Innovation, Critical Thinking and Problem Solving, Communication and Collaboration
>
> - Information, Media, and Technology Skills: Information Literacy, Media Literacy, ICT Literacy
>
> - Life and Career Skills: Flexibility and Adaptability, Initiative and Self-Direction, Social and Cross-Cultural Skills, Productivity and Accountability, Leadership and Responsibility

Table 4.14. Key Vocabulary for Lesson 5

Key Vocabulary	Definition
prototype	a preliminary model of a device
stakeholder	a person or group that has a particular interest in an event or issue

TEACHER BACKGROUND INFORMATION

As students take on fictional roles related to various stakeholder groups in this lesson, they will discover that they need to be able to think and act from the perspective of others with views and experiences different from their own. The challenges associated with this, and in working in teams composed of diverse individuals, require empathy, which is a concept that is increasingly being discussed in terms of engineering and technology innovation. Engineers and technology innovators are regularly charged with developing products or solutions that will address the needs of and be used by a wide diversity of individuals. Developing the best solutions requires that those doing the designing be able to experience others' needs in personalized ways. Information for product designers and engineers can be gained, for example, through focus groups and consumer research.

In this lesson, students will be challenged to think from perspectives different from their own without the benefit of hearing firsthand from the stakeholders whose role they are assuming. To support students' ability to empathize with the group they are representing, you may wish to have them consider questions such as the following:

- What is the focus or agenda of this stakeholder group?

- What service does this stakeholder group provide to society?

- What kind of jobs or career backgrounds do people within this stakeholder group have?

- Where does the funding for this group's work come from?

- Do you know anyone who is associated with this group? If so, how do you think they would react to a crisis like the Gammatown Crisis Challenge?

COMMON MISCONCEPTIONS

Students will have various types of prior knowledge about the concepts introduced in this lesson. Table 4.15 (p. 148) outlines some common misconceptions students may have concerning these concepts. Because of the breadth of students' experiences, it is not possible to anticipate every misconception that students may bring as they approach this lesson. Incorrect or inaccurate prior understanding of concepts can influence student learning in the future, however, so it is important to be alert to misconceptions such as those presented in the table.

Table 4.15. Common Misconceptions About the Concepts in Lesson 5

Topic	Student Misconception	Explanation
Physical models	Models are art projects.	Models are used to demonstrate and explain concepts that may be difficult to describe using only words. They may contain artistic elements, but their purpose is more than artistic.
	Models need to show every part of the object they represent.	Models should show the major features of the object they represent but do not need to include every detail.
	Models cannot be changed once they are constructed.	Using the EDP to construct a model means that the model can and should be changed and improved so that it does a better job of demonstrating and explaining the function of the object it represents.

PREPARATION FOR LESSON 5

Review the Teacher Background Information section (p. 147), assemble materials for the lesson, and make copies of the student handouts. The challenge synthesizes student learning from various content areas and activities and you should decide whether it will be addressed within science class or divided among classes. Student teams who choose to create physical models should put together their own materials lists and determine which materials are available in the classroom and which they must provide themselves. Prepare invitations for outside guests for the town hall meeting. You may wish to provide guests with background information about the challenge preparation, including the Gammatown Crisis Challenge Narrative and a list of the stakeholder groups that teams will represent. Encourage guests to ask questions after team presentations.

LEARNING COMPONENTS
Introductory Activity/Engagement

Connection to the Challenge: Begin this lesson by telling students that they will now work on creating a solution to the Gammatown Crisis Challenge. This solution will address the module's driving question: How does the use of nuclear energy to meet our energy demands affect society? Students should use the EDP to structure their work during the challenge and track their progress in their STEM Research Notebooks. Since various components and deliverables are associated with the challenge, the planning phase

of the EDP is particularly important for students. Students should create a schedule and a plan for completing the work within the designated time.

Science Class: As a class, read aloud the Gammatown Crisis Challenge Narrative (attached at the end of Lesson 1 on pp. 73–74) once again. Then, review the requirements of the challenge on the Gammatown Crisis Challenge Overview student handout (p. 72).

Mathematics Class: Not applicable.

ELA and Social Studies Connections: Not applicable.

Activity/Exploration

Science Class: Student teams create solutions to the module's final challenge, the Gammatown Crisis Challenge.

The Gammatown Crisis Challenge

Assign each student team the role of a stakeholder group. Tell students that each team will receive a handout specific to their stakeholder group, which will include some additional information about the Gammatown nuclear plant accident. Other teams may or may not have the same information, and it is up to each team whether it shares its information with the others. The information the teams receive may or may not be useful in creating their challenge solutions, so students need to consider carefully how or whether to include information.

Distribute copies of the following (attached at the end of this lesson unless otherwise noted) to each student:

- Stakeholder Group Information handout for appropriate team stakeholder group

- Stakeholder Group Presentation Rubric

- Stakeholder Group Printed Materials Rubric

- Stakeholder Group Prototype Design Rubric

- Collaboration Rubric (attached at the end of Lesson 2 on pp. 101–102)

Tell students that they should provide evidence of their team's use of the EDP in their STEM Research Notebooks, labeling a page with each step of the EDP and adding information appropriate to that step. These entries will be made in lieu of STEM Research Notebook prompts for this lesson. You may wish to provide students with a general outline for organizing this information in their notebooks, such as the following:

1. Define
   a. Identify your group's target audience.

  b. What is the goal of your presentation? (e.g., to provide information to your audience, to persuade the audience of something, to minimize people's fears, to protect the public)

  c. What products do you need to produce?

2. Learn

  a. What additional information do you need?

  b. What did you find out from your research? Be sure to provide citations for your information.

  c. What ideas do team members have?

3. Plan

  a. How will you schedule your work to ensure that you complete it on time?

  b. How will you divide tasks? (Hint: You might want to create a chart assigning team members jobs.)

  c. What will your prototype be? Make a sketch!

  d. What materials do you need?

4. Try

  a. Create the components of your response

   i. Presentation
   ii. Printed materials
   iii. Prototype

5. Test

  a. Practice your presentation and get feedback from others if possible. Make sure your audience understands your goal!

  b. What worked well?

  c. What didn't work well?

6. Decide

  a. Based on your test run, what will you change?

7. Share

  a. Share your challenge response in the town hall meeting. Make sure you have determined who will present various parts of your presentation.

After students have completed their work, hold a town hall meeting in which each team presents its challenge solution. You may wish to assign audience members various stakeholder roles and encourage them to ask questions relevant to their stakeholder group.

Mathematics Class: Students should incorporate mathematical concepts into their responses. You may wish to review the various mathematics concepts introduced over the course of the module and work with individual teams to identify the most appropriate mathematical concepts or models for them to use.

Social Studies Connection: Students may choose to incorporate information about the Environmental Protection Agency, the Nuclear Regulatory Commission (NRC), historical nuclear events and accidents, and regulations governing nuclear power plants in their challenge solutions.

ELA Connection: Not applicable.

Explanation

Science Class: Students may have additional questions about the Gammatown Crisis while completing their challenge work. You should feel free to give appropriate responses to these questions, providing additional information that is consistent with the facts provided in the Gammatown Crisis Challenge Narrative. You may wish to use the Three Mile Island account of events as a basis for your answers.

Mathematics Class: Students should incorporate mathematical concepts from the module into their final challenge solution. These must be part of the presentation and may be included in the printed materials and prototype as well.

Social Studies Connection: Students should be aware that nuclear power plant regulations have proliferated in the years since the Three Mile Island accident. They should work under the assumption that the Gammatown Crisis occurred in the current regulatory environment. While a knowledge of nuclear regulations is not necessary, students may find useful information the NRC website at *www.nrc.gov/reactors.html.*

ELA Connection: Students should incorporate persuasive language in their presentations. Review elements of persuasive language and create a class list of good presentation skills.

Elaboration/Application of Knowledge

Science Class: Student teams should present their challenge solutions and prototypes to guests assembled for the town hall meeting. Allow time for guests to ask questions after each team's presentation.

Mathematics Class: Not applicable.

ELA and Social Studies Connections: Not applicable.

Evaluation/Assessment

Students may be assessed on the following performance tasks and other measures listed.

Performance Tasks

- Stakeholder Group Presentation Rubric (pp. 160–162)

- Stakeholder Group Printed Materials Rubric (p. 163)

- Stakeholder Group Prototype Design Rubric (p. 164)

Other Measures

- STEM Research Notebook entries on the EDP

- Collaboration Rubric (attached at the end of Lesson 2 on pp. 101–102)

INTERNET RESOURCE

NRC information about nuclear reactors
- *www.nrc.gov/reactors.html*

STUDENT HANDOUT

STAKEHOLDER GROUP INFORMATION

Public Health Officials

Public health is generally above average in Gammatown. For the past five years, your office has been running a campaign for blood pressure awareness that includes encouraging people to consume less salt. In fact, the local diner no longer salts the french fries it serves, and the tabletop salt shakers contain a nonsodium salt substitute. Your records show that there has been an overall decrease in blood pressure, indicating that the campaign has been a success. At the same time, however, there has been a recent uptick in the number of cases of thyroid disease (in particular, hypothyroidism) in Gammatown. In fact, just last month, before the accident, the Gammatown Public Health Office reported this increase in thyroid disorders to the Centers for Disease Control, and the local paper published an article about the report. Your office has been receiving calls asking whether this could have been due to radiation leaks from the Gammatown Power Plant before the accident.

Rumor and misinformation seem to be rampant in Gammatown. People have been sealing their windows and doors with plastic and duct tape, buying bottled water, and wearing surgical masks when walking through town. Puddles of bright green liquid have been spotted around town. Your office suspects that this might be the work of practical jokers, since your lab tests have shown that this liquid is Gatorade. However, you have not been successful in getting this news to the public, and the puddles have caused a mild panic.

Your team must address citizens' general health concerns and lack of knowledge about radiation exposure, as well as the public's concern about the increasing numbers of thyroid disease cases.

STUDENT HANDOUT

STAKEHOLDER GROUP INFORMATION

Power Plant Public Relations Team

Your team is overworked and understaffed. You are contending with hundreds of reporters from the regional, national, and international press who have descended on Gammatown since the accident. These journalists have been combing the power plant's records, which are publicly available because this is a public utility. They discovered that five years ago, recurring equipment problems led to an emergency shutdown of reactor #2. This was a safety precaution, and no radiation was released into the environment or reactor building, but reporters are sure you're hiding something. You tried to hold a press conference, but it was unproductive, since the reporters don't understand how a nuclear reactor works and kept asking when it will explode. You have also been getting questions about puddles of bright green liquid that have been found around town. You have no idea what they are talking about and assume that this is just a rumor.

The chief relations officer finally called a meeting to devise a strategy to communicate with the press. He has scheduled a press conference and received a commitment from most members of the press to listen carefully to what your team tells them. Most of the people at the press conference will have almost no background in science and no understanding of how a nuclear power plant operates; however, they will be responsible for conveying important information to the public in the coming weeks. It is your job to inform them so they are able to do this knowledgeably.

STUDENT HANDOUT

STAKEHOLDER GROUP INFORMATION

Power Plant Operators

The recent accident at the Gammatown Power Plant has attracted a great deal of unwelcome attention from the Nuclear Regulatory Commission (NRC). The plant's operating license is set to expire in 2020, but you hope to have the license extended for years into the future. The plant has a strong safety record overall; however, an NRC inspection last month showed that four fire extinguishers in the vicinity of the control room and the reactor building were out of date. In addition, there is ongoing concern about flooding because of the plant's location near the Scenic River and the recent accident, and the plant's future is looking bleak.

Your team must convince the NRC that the Gammatown plant provides a valuable service to the community with little risk to the public. NRC representatives will attend the town hall meeting, so you can present your information to them there. Your task is to highlight all the safety precautions that have been followed at the plant over the years and the preventive and remedial actions that were taken when the malfunction was identified on June 2. You also must assure your audience that the plant will be even safer in the future.

The Gammatown incident has received so much attention that members of Congress have requested that your presentation to the NRC also include any U.S. senators and representatives who wish to attend. This complicates your task, since these individuals have a great deal of influence but know little about the science of nuclear energy or nuclear power plant operation.

STUDENT HANDOUT

STAKEHOLDER GROUP INFORMATION

Nuclear Regulatory Commission (NRC) Officials

Your team is in a tricky position. The Gammatown plant has a good safety record overall, but it is unlikely that this will mean much to the president and Congress, given recent events. You know that the president and members of Congress have been studying the Three Mile Island accident, since they know that this accident was ultimately found to be much worse than originally thought. (Years later, it was discovered that much of the core had become uncovered, unbeknownst to anyone at the time.) The fear is that the Gammatown incident is similar to the Three Mile Island accident, and the nation's leaders are concerned that early assurances that the accident is under control will prove to be false.

Your team has decided that the best plan is to compare the Gammatown incident with the Three Mile Island accident, highlighting the points where the Gammatown operators made different—and better—decisions. (You may assume that the Gammatown operators made the right decisions at key points where the Three Mile Island operators made poor decisions or took actions based on faulty information.)

You will present your materials at a Gammatown town hall meeting. You have been informed that the president and members of Congress will be watching the meeting remotely, and it is possible that some of them might even attend the meeting. Since your audience does not have an understanding of nuclear science or nuclear power plant operations, your job is to create simple yet informative and accurate materials.

STUDENT HANDOUT

STAKEHOLDER GROUP INFORMATION

Environmental Protection Agency (EPA) Team

Your team has been dispatched to Gammatown as part of the EPA's Radiological Emergency Response Team. You have set up a mobile environmental radiation laboratory, where you provide radiation analysis of samples. Your task has become overwhelming lately because of the large number of samples provided by the public. People have begun bringing fur from their cats and dogs and strands of their children's hair for testing, and one woman even brought your team a jar full of a bright green liquid she claims she found in a puddle on the street. To avoid chaos and make sure that your tests reflect environmental impact, your team will now test only samples that it collects. To date, the highest radiation content your team tested from the air around the plant was 35 picocuries per liter. The public is worried about reports of iodine-131 and caesium-137 release into the air around the plant. Many people are also frightened of the Geiger counters you use and believe that they may actually spread radioactivity into the environment.

Your job is to report on your role, your work, and your findings to the public and to local government officials. You should include specific information about how a Geiger counter works.

STUDENT HANDOUT

STAKEHOLDER GROUP INFORMATION

Antinuclear Activists

You are a part of the No Nukes Nation group, an antinuclear group that opposes any use of nuclear energy for energy production or weaponry. Your team deployed to Gammatown as soon as you heard about the accident and has been vying for media attention ever since. Your task is to incite fear of nuclear energy into residents of Gammatown while sticking to the facts. Your group's modus operandi is to include information on the science of nuclear fission while highlighting the harmful effects of past accidents. In this way, you hope to garner public support to shut down the Gammatown plant for good. Although no individual in your group will admit to it, you suspect that a No Nukes Nation team member may be responsible for the bright green puddles around town that have caused some panic, since some team members have been known to resort to drastic scare tactics in support of the cause.

Your team's job is to lead a public campaign to garner support for your cause and opposition to the Gammatown plant and nuclear power in general.

STUDENT HANDOUT

STAKEHOLDER GROUP INFORMATION

Business Owners and Economic Development Officials

The news about the Gammatown reactor accident couldn't have come at a worse time for your group. Seat Belts R Us is on the brink of committing to build a Gammatown seat belt factory that will put your town on the map. Since the plant will employ hundreds of people, not only will local residents find well-paying jobs, but people will move from other areas to Gammatown, increasing the tax base and supporting area business owners. Right after the accident, you got a call from the president of Seat Belts R Us, Rick Restraint, expressing hesitancy about the deal. He's coming to town along with a group of other executives from the company to make the final decision about whether the Gammatown project should move forward. Mr. Restraint was very clear during his phone call that he and the other executives like to have the facts, but they don't understand nuclear energy or what happened at Gammatown, and they are worried about the future livability of the town and the possibility that the air and water around Gammatown are contaminated with radioactivity.

Your job is to convince Mr. Restraint and the other seat belt executives that Gammatown is a safe place to work and live. You are particularly worried about the influence of No Nukes Nation, a group of antinuclear activists who established a presence in town right after the accident, since you know that their goal is to cast the accident in the worst possible light.

Stakeholder Group Presentation Rubric

Team Name: _____

Team Performance	Needs Improvement (1–3 points)	Approaches Expectations (4–6 points)	Meets or Exceeds Expectations (7–9 points)	Team Score
GENERAL CONTENT	• Information is not appropriate to stakeholder group audience; it may be unclear what stakeholder group the team represents. • Includes incorrect or misleading information. • Few if any facts are included.	• Information is somewhat appropriate to stakeholder group. • Includes some incorrect or misleading information. • Some facts are included.	• Information is appropriate to stakeholder group. • Information is correct and relevant. • Appropriate and relevant facts are included to support the team's position.	
SCIENTIFIC CONTENT	• Science content is inaccurate or incomplete. • There is little evidence that students understand nuclear fission and its use in nuclear power plants.	• Science content is included, but it may contain inaccuracies. • Students provide some evidence of understanding nuclear fission and its use in nuclear power plants.	• Complete and accurate science content is included. • Students demonstrate an understanding of nuclear fission and its use in nuclear power plants.	
MATHEMATICS CONTENT	• Little or no mathematics content is included. • Mathematics content is inaccurate or inappropriate to topic.	• Mathematics content is included, although it may be only marginally related to the topic and may contain inaccuracies.	• Accurate and relevant mathematical concepts are incorporated into the team's presentation.	
SOURCES OF INFORMATION	• Team does not include references to information sources.	• Team includes some references to sources of information.	• Team includes multiple complete sources for research.	

Continued

Stakeholder Group Presentation Rubric (*continued*)

Team *Performance*	Needs Improvement (1–3 points)	Approaches Expectations (4–6 points)	Meets or Exceeds Expectations (7–9 points)	Team *Score*
IDEAS AND ORGANIZATION	• Team does not have a main idea or organizational strategy. • Presentation does not include an introduction or a conclusion. • Presentation is confusing and uninformative. • Team uses presentation time poorly.	• Team has a main idea or organizational strategy, but it is not clear or coherent. • Presentation includes either an introduction or a conclusion, but not both. • Presentation is somewhat coherent but is not well organized and is only somewhat informative. • Presentation may be somewhat too long or too short.	• Team has a clear main idea and organizational strategy. • Presentation includes both an introduction and a conclusion. • Presentation is coherent, well organized, and informative. • Team uses presentation time well, and presentation is neither too short nor too long.	
PRESENTATION STYLE	• Only one or two team members participate in the presentation. • Presenters are difficult to understand. • Presentation is not creative and not interesting to watch. • Presenters use language inappropriate for audience (e.g., slang, poor grammar, frequent filler words such as "uh," "um").	• Some, but not all, team members participate in the presentation. • Most presenters are understandable, but volume may be too low or some presenters may mumble. • Presentation has some creativity. • Presenters use some language inappropriate for audience (e.g., slang, poor grammar, some use of filler words such as "uh," "um").	• All team members participate in the presentation. • Presenters are easy to understand. • Presentation displays creativity and is interesting to watch. • Presenters use appropriate language for audience (no slang, poor grammar, infrequent use of filler words such as "uh," "um").	

Continued

Stakeholder Group Presentation Rubric (*continued*)

Team Performance	Needs Improvement (1–3 points)	Approaches Expectations (4–6 points)	Meets or Exceeds Expectations (7–9 points)	Team Score
VISUAL AIDS	• Team does not use any visual aids in presentation.	• Team uses some visual aids in presentation, but they may be poorly executed or detract from the presentation.	• Team uses well-produced visual aids or media that clarify and enhance presentation.	
RESPONSE TO AUDIENCE QUESTIONS	• Team fails to respond to questions from audience or responds inappropriately.	• Team responds appropriately to audience questions, but responses may be brief, incomplete, or unclear.	• Team responds clearly and in detail to audience questions and seeks clarification of questions when necessary.	

TOTAL SCORE: _____

COMMENTS:

Team Name: _____

Stakeholder Group Printed Materials Rubric

Team Performance	Needs Improvement (1–3 points)	Approaches Expectations (4–6 points)	Meets or Exceeds Expectations (7–9 points)	Team Score
GENERAL CONTENT	• Information is not appropriate to stakeholder group; it may be unclear what stakeholder group the team represents. • Includes incorrect or misleading information. • Few, if any, facts are included.	• Information is somewhat appropriate to stakeholder group. • Includes some incorrect or misleading information. • Some facts are included.	• Information is appropriate to stakeholder group. • Information is correct and relevant. • Appropriate and relevant facts are included to support the team's position.	
AUDIENCE APPROPRIATENESS	• Content or format is inappropriate for team's target audience. • Content is not informative or is inaccurate.	• Content is appropriate for team's target audience, but format may not be appropriate or accessible for this audience. • Content is informative and mostly accurate.	• Content and format are clearly appropriate for team's target audience. • Content is informative and accurate.	
SOURCES OF INFORMATION	• Team does not include references to information sources.	• Team includes some references to sources of information.	• Team includes multiple sources for research. • Team includes complete references.	
APPEARANCE	• Materials are not neatly constructed or are difficult to read and understand.	• Materials are neat and readable.	• Materials are neat, visually appealing, and readable.	

TOTAL SCORE: _____

COMMENTS:

Stakeholder Group Prototype Design Rubric

Team Name: _____

Team Performance	Needs Improvement (1–3 points)	Approaches Expectations (4–6 points)	Meets or Exceeds Expectations (7–9 points)	Team Score
CREATIVITY AND INNOVATION	• Design reflects little creativity with use of materials or concepts. • Design is impractical. • Design has several elements that do not fit.	• Design reflects some creativity with use of materials or concepts, a basic understanding of project purpose, and limited innovative design features. • Design is limited in practicality and function. • Design has some interesting elements.	• Design reflects creative use of materials and concepts and a sound understanding of project purpose. • Design is practical. • Design includes interesting elements that are appropriate for the purpose.	
CONCEPTUAL UNDERSTANDING	• Design incorporates no or few features that reflect conceptual understanding of the science concepts in the unit.	• Design incorporates some features that reflect a limited conceptual understanding of science concepts.	• Design incorporates several features that reflect a sound conceptual understanding of science concepts.	
DESIGNED WITHIN SPECIFIED REQUIREMENTS	• Design is not finished or provides little to no value to the team's challenge solution.	• Design is finished on time and adds some value to the team's challenge solution.	• Design is finished on time and clearly adds value to the team's challenge solution.	

TOTAL SCORE: _____

COMMENTS:

TRANSFORMING LEARNING WITH RADIOACTIVITY AND THE *STEM ROAD MAP CURRICULUM SERIES*

Carla C. Johnson

This chapter serves as a conclusion to the Radioactivity integrated STEM curriculum module, but it is just the beginning of the transformation of your classroom that is possible through use of the *STEM Road Map Curriculum Series.* In this book, many key resources have been provided to make learning meaningful for your students through integration of science, technology, engineering, and mathematics, as well as social studies and English language arts, into powerful problem- and project-based instruction. First, the Radioactivity curriculum is grounded in the latest theory of learning for students in grade 11 specifically. Second, as your students work through this module, they engage in using the engineering design process (EDP) and build prototypes like engineers and STEM professionals in the real world. Third, students acquire important knowledge and skills grounded in national academic standards in mathematics, English language arts, science, and 21st century skills that will enable their learning to be deeper, retained longer, and applied throughout, illustrating the critical connections within and across disciplines. Finally, authentic formative assessments, including strategies for differentiation and addressing misconceptions, are embedded within the curriculum activities.

The Radioactivity curriculum in The Represented World STEM Road Map theme can be used in single-content classrooms where there is only one teacher or expanded to include multiple teachers and content areas across classrooms. Through the exploration of the Gammatown Crisis Challenge, students engage in a real-world STEM problem on the first day of instruction and gather necessary knowledge and skills along the way in the context of solving the problem.

The other topics in the *STEM Road Map Curriculum Series* are designed in a similar manner, and NSTA Press has additional volumes in this series for this and other grade levels and plans to publish more. The volumes covering Innovation and Progress have been published and are as follows:

- *Amusement Park of the Future, Grade 6*

- *Construction Materials, Grade 11*

- *Harnessing Solar Energy, Grade 4*

- *Transportation in the Future, Grade 3*

- *Wind Energy, Grade 5*

In addition, the other volumes covering The Represented World have been published:

- *Car Crashes, Grade 12*

- *Improving Bridge Design, Grade 8*

- *Investigating Environmental Changes, Grade 2*

- *Packaging Design, Grade 6*

- *Patterns and the Plant World, Grade 1*

- *Rainwater Analysis, Grade 5*

- *Swing Set Makeover, Grade 3*

The tentative list of other books includes the following themes and subjects:

- Cause and Effect

  - Influence of waves

  - Hazards and the changing environment

  - The role of physics in motion

- Sustainable Systems

  - Creating global bonds

  - Composting: Reduce, reuse, recycle

  - Hydropower efficiency

  - System interactions

- Optimizing the Human Experience

  - Genetically modified organisms

  - Mineral resources

  - Rebuilding the natural environment

  - Water conservation: Think global, act local

If you are interested in professional development opportunities focused on the STEM Road Map specifically or integrated STEM or STEM programs and schools overall, contact the lead editor of this project, Dr. Carla C. Johnson (*carlacjohnson@ncsu.edu*), associate dean and professor of science education and executive director of the William and Ida Friday Institute at North Carolina State University. Someone from the team will be in touch to design a program that will meet your individual, school, or district needs.

APPENDIX

CONTENT STANDARDS ADDRESSED IN THIS MODULE

NEXT GENERATION SCIENCE STANDARDS

Table A1 (p. 170) lists the science and engineering practices, disciplinary core ideas, and crosscutting concepts this module addresses. The supported performance expectations are as follows:

- HS-PS1-1. Use the periodic table as a model to predict the relative properties of elements based on the patterns of electrons in the outermost energy level of atoms.

- HS-PS1-7. Use mathematical representations to support the claim that atoms, and therefore mass, are conserved during a chemical reaction.

- HS-PS1-8. Develop models to illustrate the changes in the composition of the nucleus of the atom and the energy released during the processes of fission, fusion, and radioactive decay.

- HS-ETS-2. Design a solution to a complex real-world problem by breaking it down into smaller, more manageable problems that can be solved through engineering.

Table A1. *Next Generation Science Standards (NGSS)*

Science and Engineering Practices

USING MATHEMATICS AND COMPUTATIONAL THINKING

- Use mathematical representations of phenomena to support claims.

- Create a computational model or simulation of a phenomenon, designed device, process, or system.

DEVELOPING AND USING MODELS

- Use a model to predict the relationships between systems or between components of a system.

- Develop a model based on evidence to illustrate the relationships between systems or between components of a system.

OBTAINING AND COMMUNICATING INFORMATION

- Evaluate the validity and reliability of multiple claims that appear in scientific and technical texts or media reports, verifying the data when possible.

- Communicate technical information or ideas in multiple formats.

CONSTRUCTING EXPLANATIONS AND DESIGNING SOLUTIONS

- Design, evaluate, and/or refine a solution to a complex real-world problem, based on scientific knowledge, student-generated sources of evidence, prioritized criteria, and trade-off considerations.

- Construct and revise an explanation based on valid and reliable evidence obtained from a variety of sources (including students' own investigations, models, theories, simulations, peer review) and the assumption that theories and laws that describe the natural world operate today as they did in the past and will continue to do so in the future.

Disciplinary Core Ideas

PS1.A. STRUCTURE AND PROPERTIES OF MATTER

- Each atom has a charged substructure consisting of a nucleus, which is made of protons and neutrons, surrounded by electrons.

- The periodic table orders elements horizontally by the number of protons in the atom's nucleus and places those with similar chemical properties in columns. The repeating patterns of this table reflect patterns of outer electron states.

- A stable molecule has less energy than the same set of atoms separated; one must provide at least this energy in order to take the molecule apart.

PS1.B. CHEMICAL REACTIONS

- The fact that atoms are conserved, together with knowledge of the chemical properties of the elements involved, can be used to describe and predict chemical reactions.

Continued

Table A1. (*continued*)

PS1.C. NUCLEAR PROCESSES

- Nuclear processes, including fusion, fission, and radioactive decays of unstable nuclei, involve release or absorption of energy. The total number of neutrons plus protons does not change in any nuclear process.

- Spontaneous radioactive decays follow a characteristic exponential decay law. Nuclear lifetimes allow radiometric dating to be used to determine the ages of rocks and other materials.

PS2.B. TYPES OF INTERACTIONS

- Attraction and repulsion between electric charges at the atomic scale explain the structure, properties, and transformations of matter, as well as the contact forces between material objects.

PS3.A. DEFINITIONS OF ENERGY

- Energy is a quantitative property of a system that depends on the motion and interactions of matter and radiation within that system. That there is a single quantity called energy is due to the fact that a system's total energy is conserved, even as, within the system, energy is continually transferred from one object to another and between its various possible forms.

PS3.B. CONSERVATION OF ENERGY AND ENERGY TRANSFER

- Conservation of energy means that the total change of energy in any system is always equal to the total energy transferred into or out of the system.

- Energy cannot be created or destroyed, but it can be transported from one place to another and transferred between systems.

- Mathematical expressions, which quantify how the stored energy in a system depends on its configuration (e.g., relative positions of charged particles, compression of a spring) and how kinetic energy depends on mass and speed, allow the concept of conservation of energy to be used to predict and describe system behavior.

- The availability of energy limits what can occur in any system.

PS3.D. ENERGY IN CHEMICAL PROCESSES

- Although energy cannot be destroyed, it can be converted to less useful forms—for example, to thermal energy in the surrounding environment.

- Nuclear fusion processes in the center of the sun release the energy that ultimately reaches Earth as radiation.

ETS1.A. DEFINING AND DELIMITING AN ENGINEERING PROBLEM

- Criteria and constraints also include satisfying any requirements set by society, such as taking issues of risk mitigation into account, and they should be quantified to the extent possible and stated in such a way that one can tell if a given design meets them.

Continued

Table A1. (*continued*)

ETS1.C. OPTIMIZING THE DESIGN SOLUTION
• Criteria may need to be broken down into simpler ones that can be approached systematically, and decisions about the priority of certain criteria over others (trade-offs) may be needed.

CROSSCUTTING CONCEPTS

PATTERNS
- Different patterns may be observed at each of the scales at which a system is studied and can provide evidence for causality in explanations of phenomena.
- Empirical evidence is needed to identify patterns.

ENERGY AND MATTER
- Changes of energy and matter in a system can be described in terms of energy and matter flows into, out of, and within that system.
- In nuclear processes, atoms are not conserved, but the total number of protons plus neutrons is conserved.
- The total amount of energy and matter in closed systems is conserved.

STABILITY AND CHANGE
- Much of science deals with constructing explanations of how things change and how they remain stable.

CAUSE AND EFFECT
- Cause and effect relationships can be suggested and predicted for complex natural and human designed systems by examining what is known about smaller scale mechanisms within the system.

SCALE, PROPORTION, AND QUANTITY
- The significance of a phenomenon is dependent on the scale, proportion, and quantity at which it occurs.

SYSTEMS AND SYSTEM MODELS
- Models can be used to predict the behavior of a system, but these predictions have limited precision and reliability due to the assumptions and approximations inherent in models.

STRUCTURE AND FUNCTION
- Investigating or designing new systems or structures requires a detailed examination of the properties of different materials, the structures of different components, and connections of components to reveal its function and/or solve a problem.

Source: NGSS Lead States. 2013. *Next Generation Science Standards: For states, by states.* Washington, DC: National Academies Press. *www.nextgenscience.org/next-generation-science-standards.*

Table A2. Common Core Mathematics and English Language Arts (ELA) Standards

<table>
<tr><td>

MATHEMATICAL PRACTICES

- MP1. Make sense of problems and persevere in solving them.
- MP3. Construct viable arguments and critique the reasoning of others.
- MP4. Model with mathematics.
- MP7. Look for and make use of structure.

MATHEMATICAL CONTENT

- HSA.APR.D.6. Rewrite simple rational expressions in different forms; write $a(x)/b(x)$ in the form $q(x) + r(x)/b(x)$, where $a(x)$, $b(x)$, $q(x)$, and $r(x)$ are polynomials with the degree of $r(x)$ less than the degree of $b(x)$ using inspection, long division, or, for the more complicated examples, a computer algebra system.
- HSF.IF.B.4. For a function that models a relationship between two quantities, interpret key features of graphs and tables in terms of the quantities, and sketch graphs showing key features given a verbal description of the relationship.
- HSF.IF.B.6. Calculate and interpret the average rate of change of a function (presented symbolically or as a table) over a specified interval. Estimate the rate of change from a graph.
- HSF.LE.A.1. Distinguish between situations that can be modeled with linear functions and with exponential functions.
- HSF.LE.A.2. Construct linear and exponential functions, including arithmetic and geometric sequences, given a graph, a description of a relationship, or two input-output pairs (include reading these from a table).
- HSF.LE.A.3. Observe using graphs and tables that a quantity increasing exponentially eventually exceeds a quantity increasing linearly, quadratically, or (more generally) as a polynomial function.
- HSF.LE.A.4. For exponential models, express as a logarithm the solution to $ab^{ct} = d$ where a, c, and d are numbers and the base b is 2, 10, or e; evaluate the logarithm using technology.

</td><td>

READING STANDARDS

- RI.11-12.1. Cite strong and thorough textual evidence to support analysis of what the text says explicitly as well as inferences drawn from the text, including determining where the text leaves matters uncertain.
- RI.11-12.2. Determine two or more central ideas of a text and analyze their development over the course of the text, including how they interact and build on one another to provide a complex analysis; provide an objective summary of the text.
- RI.11-12.3. Analyze a complex set of ideas or sequence of events and explain how specific individuals, ideas, or events interact and develop over the course of the text.
- RI.11-12.4. Determine the meaning of words and phrases as they are used in a text, including figurative, connotative, and technical meanings; analyze how an author uses and refines the meaning of a key term or terms over the course of a text.
- RI.11-12.5. Analyze and evaluate the effectiveness of the structure an author uses in his or her exposition or argument, including whether the structure makes points clear, convincing, and engaging.
- RI.11-12.7. Integrate and evaluate multiple sources of information presented in different media or formats (e.g., visually, quantitatively) as well as in words in order to address a question or solve a problem.
- RI.11-12.8. Delineate and evaluate the reasoning in seminal U.S. texts, including the application of constitutional principles and use of legal reasoning and the premises, purposes, and arguments in works of public advocacy.

WRITING STANDARDS

- W.11-12.1. Write arguments to support claims in an analysis of substantive topics or texts, using valid reasoning and relevant and sufficient evidence.
- W.11-12.2. Write informative/explanatory texts to examine and convey complex ideas, concepts, and information clearly and accurately through the effective selection, organization, and analysis of content.

</td></tr>
</table>

Coantinued

Table A2. (*continued*)

MATHEMATICAL CONTENT (*continued*)	SPEAKING AND LISTENING STANDARDS
• HSF.LE.B.5. Interpret the parameters in a linear or exponential function in terms of a context. • HSF.BF.A.1. Write a function that describes a relationship between two quantities. • HSF.BF.A.2. Write arithmetic and geometric sequences both recursively and with an explicit formula, use them to model situations, and translate between the two forms. • HSF.BF.B.3. Identify the effect on the graph of replacing $f(x)$ by $f(x) + k$, $k\,f(x)$, $f(kx)$, and $f(x + k)$ for specific values of k (both positive and negative); find the value of k given the graphs. Experiment with cases and illustrate an explanation of the effects on the graph using technology. Include recognizing even and odd functions from their graphs and algebraic expressions for them. • HSF.BF.B.4. Find inverse functions. • HSF.BF.B.5. Understand the inverse relationship between exponents and logarithms and use this relationship to solve problems involving logarithms and exponents.	• SL.11-12.1. Initiate and participate effectively in a range of collaborative discussions (one-on-one, in groups, and teacher-led) with diverse partners on grades 11–12 topics, texts, and issues, building on others' ideas and expressing their own clearly and persuasively. • SL.11-12.2. Integrate multiple sources of information presented in diverse formats and media (e.g., visually, quantitatively, orally) in order to make informed decisions and solve problems, evaluating the credibility and accuracy of each source and noting any discrepancies among the data. • SL.11-12.3. Evaluate a speaker's point of view, reasoning, and use of evidence and rhetoric, assessing the stance, premises, links among ideas, word choice, points of emphasis, and tone used. • SL.11-12.4. Present information, findings, and supporting evidence, conveying a clear and distinct perspective, such that listeners can follow the line of reasoning, alternative or opposing perspectives are addressed, and the organization, development, substance, and style are appropriate to purpose, audience, and a range of formal and informal tasks. • SL.11-12.5. Make strategic use of digital media (e.g., textual, graphical, audio, visual, and interactive elements) in presentations to enhance understanding of findings, reasoning, and evidence and to add interest. • SL.11-12.6. Adapt speech to a variety of contexts and tasks, demonstrating a command of formal English when indicated or appropriate.

Source: National Governors Association Center for Best Practices and Council of Chief State School Officers (NGAC and CCSSO). 2010. *Common core state standards.* Washington, DC: NGAC and CCSSO.

Table A3. 21st Century Skills From the Framework for 21st Century Learning

21st Century Skills	Learning Skills and Technology Tools	Teaching Strategies	Evidence of Success
INTERDISCIPLINARY THEMES • Global Awareness • Financial, Economic, Business and Entrepreneurial Literacy • Civic Literacy • Environmental Literacy	• Using 21st century skills to understand and engage in discourse about global issues. • Demonstrate knowledge and understanding of society's impact on the natural world. • Investigate and analyze environmental issues and draw accurate conclusions about effective solutions.	• Emphasize both the positive and negative impacts of scientific and technical advances on society, business, and the world ecosystem.	• Students can to discuss the positive and negative impacts of nuclear energy on society.
LEARNING AND INNOVATION SKILLS • Creativity and Innovation • Critical Thinking and Problem Solving • Communication and Collaboration	• Use a wide range of idea creation techniques (such as brainstorming) to create new ideas and refine and evaluate those ideas. • Be able to view issues through the lens of new and diverse perspectives. • Demonstrate originality and inventiveness in work and understanding the real world limits to adopting new ideas. • Analyze how parts of a complex system interact with one another to produce overall outcomes. • Effectively analyze and evaluate evidence, arguments, claims, and beliefs and draw conclusions.	• Guide students to resources outlining the history of nuclear energy and research. • Incorporate the concepts of half-life and chain reactions into activities requiring problem solving and critical thinking.	• Students have independently and collaboratively synthesized information they learned about nuclear decay and fission and the implications of nuclear power and used this understanding to address the Gammatown Crisis Challenge.

Continued

Table A3. (*continued*)

21st Century Skills	Learning Skills and Technology Tools	Teaching Strategies	Evidence of Success
	• Articulate thoughts and ideas effectively using oral, written, and nonverbal communication skills in a variety of forms and contexts. • Listen effectively to decipher meaning, including knowledge, values, attitudes, and intentions. • Use communication for a range of purposes (e.g., to inform, instruct, motivate, and persuade). • Demonstrate ability to work effectively and respectfully with diverse teams. • Exercise flexibility and willingness to be helpful in making necessary compromises to accomplish a common goal. • Assume shared responsibility for collaborative work, and value the individual contributions made by each team member.		
INFORMATION, MEDIA, AND TECHNOLOGY SKILLS • Information Literacy • Media Literacy • ICT Literacy	• Access information efficiently (time) and effectively (sources). • Evaluate information critically and competently. • Use information accurately and creatively for the issue or problem at hand. • Understand both how and why media messages are constructed, and for what purposes.	• Require the use of multiple reliable sources for obtaining information about nuclear energy.	• Students used a variety of reliable resources and cited these resources appropriately.

Continued

Table A3. (*continued*)

21st Century Skills	Learning Skills and Technology Tools	Teaching Strategies	Evidence of Success
LIFE AND CAREER SKILLS • Flexibility and Adaptability • Initiative and Self-Direction • Social and Cross-Cultural Skills • Productivity and Accountability • Leadership and Responsibility	• Adapt to varied roles, jobs, responsibilities, schedules, and contexts. • Incorporate feedback effectively. • Deal positively with praise, setbacks, and criticism. • Understand, negotiate, and balance diverse views and beliefs to reach workable solutions, particularly in multi-cultural environments. • Balance tactical (short-term) and strategic (long-term) goals. • Utilize time and manage workload efficiently. • Monitor, define, prioritize, and complete tasks without direct oversight. • Reflect critically on past experiences in order to inform future progress. • Know when it is appropriate to listen and when to speak. • Conduct themselves in a respectable, professional manner. • Use interpersonal and problem-solving skills to influence and guide others toward a goal. • Leverage strengths of others to accomplish a common goal.	• Provide check points for students to self-monitor their progress.	• Students articulated their goals for each check point and devised strategic plans to show progress toward their goals. • Students worked effectively in collaborative groups and were clear about roles of each member. • Students used feedback to enhance their presentations and models.

Source: Partnership for 21st Century Learning. 2015. Framework for 21st Century Learning. *www.p21.org/our-work/p21-framework.*

Table A4. English Language Development Standards

ELD STANDARD 1: SOCIAL AND INSTRUCTIONAL LANGUAGE

English language learners communicate for Social and Instructional purposes within the school setting.

ELD STANDARD 2: THE LANGUAGE OF LANGUAGE ARTS

English language learners communicate information, ideas, and concepts necessary for academic success in the content area of Language Arts.

ELD STANDARD 3: THE LANGUAGE OF MATHEMATICS

English language learners communicate information, ideas, and concepts necessary for academic success in the content area of Mathematics.

ELD STANDARD 4: THE LANGUAGE OF SCIENCE

English language learners communicate information, ideas, and concepts necessary for academic success in the content area of Science.

ELD STANDARD 5: THE LANGUAGE OF SOCIAL STUDIES

English language learners communicate information, ideas, and concepts necessary for academic success in the content area of Social Studies.

Source: WIDA, 2012. 2012 amplification of the English language development standards: Kindergarten–grade 12. *https://wida.wisc.edu/teach/standards/eld.*

INDEX

Page numbers printed in **boldface type** indicate tables, figures, or handouts.

NATIONAL SCIENCE TEACHERS ASSOCIATION